MW00843712

Lode and Placer Gold Deposits of New Mexico

"Lode and Placer Gold Deposits of New Mexico" is actually a compilation of three rare and long out of print publications. USGS Bulletin 1348 **Placer Gold Deposits of New Mexico** by Maureen G. Johnson published in 1968; US Bureau of Mines Information Circular 6987 **Gold Mining in New Mexico** by O.H. Metzger published in 1938; NM Bureau of Mines Circular Number 5 **Gold Mining and Gold Deposits in New Mexico** by E.H. Wells and T.P. Wootton published in 1932 and revised In 1957.

Miningbooks.com puts rare and valuable out of print information back into print in the field of Geology, Mining, and Metallurgy. This book gives the history of Gold Mining in New Mexico for each area where gold has been produced. Production Records, Geology of the Deposits, Locations, Districts, and information on individual mines are all covered in this book.

©2011 Miningbooks.com

Lode and Placer Gold Deposits of New Mexico

Part I. Placer Gold Deposits of New Mexico by Maureen G. Johnson, Geological Survey Bulletin 1348 – A Catalog of location, geology, and production, with lists of annotated references pertaining to the Placer Districts.

Part II. Gold Mining and Gold Deposits in New Mexico by E.H. Wells and T.P Wootton, NM Bureau of Mines & Mineral Resources Circular No. 5.

Part III. Gold Mining in New Mexico by O.H. Metzger, Bureau of Mines Information Circular 6987.

PART I

Placer Gold Deposits of New Mexico

By MAUREEN G. JOHNSON

GEOLOGICAL SURVEY BULLETIN 1348

A catalog of location, geology, and production, with lists of annotated references pertaining to the placer districts.

CONTENTS

PLACER GOLD DEPOSITS OF NEW MEXICO

By Maureen G. Johnson

ABSTRACT

Thirty-three placer districts in New Mexico are estimated to have produced a minimum of 661,000 ounces of placer gold from 1828 to 1968. The location, areal extent, past production, mining history, and probable lode source of each district are summarized from a wide variety of published reports relating to placer deposits. An annotated bibliography of all reports that give information about individual deposits is given for each district.

Most placer gold deposits in New Mexico are derived from gold-bearing mineralized areas in Tertiary intrusive rocks, and occur in gravels of alluvial fans, gulches, and rivers adjacent to the source. A few deposits are derived from gold-bearing Precambrian crystalline or Tertiary volcanic rocks. Most of the major placer districts were discovered and extensively worked between 1828 and 1880; in later years, large-scale dredging operations were successful at a few localities, while intermittent activity continued at most districts.

INTRODUCTION

HISTORY OF PLACER MINING IN NEW MEXICO

Placer mining began in 1828, when the rich placer deposits of the Ortiz Mountains, Santa Fe County, were discovered (Jones, 1904, p. 21). Even before that discovery, New Mexico was the scene of some mining activity. The Pueblo Indians mined and used turquoise for ornaments, and there is some evidence to suggest that they used gold (probably collected from gravel deposits). Alvar Nunez Cabeza de Vaca, who stayed with the Pueblo Indians on his journey from Florida to Mexico, spread stories of gold and silver possessed by these Indians. Coronado left Mexico in 1540 with a large expedition to find and exploit the source of this supposed vast wealth. He found no great riches, and, after exploring the land, returned to Mexico in 1542. The Spanish returned to the territory in the late 1500's, established missions, and may have done some mining, for after the Pueblo revolt in 1680 the Indians stipulated that the Spaniards were not to engage in mining but were to confine their activities to agriculture. Placers were reportedly worked along the Rio Grande (Taos County) in the 1600's. The real

1

development of mineral resources began in 1800 after an Indian disclosed the location of the rich Santa Rita copper deposits to a Spanish officer. The first shipments of copper to Mexico were made in 1801.

Gold was discovered in the Ortiz Mountains at Old Placers in 1828. New Placers was discovered in 1839, and before 1846 minor deposits of placer gold were found at Taos and Abiquiu and in the Sangre de Cristo Mountains (Prince, 1883, p. 243). The placers along the Rio Grande were probably mined intermittently from 1600 to 1828—placer gold was found at Rio Hondo in 1826—but it was the discovery in the Ortiz Mountains that marked the beginning of real interest in New Mexico placers. During the decade 1860–70, many placer deposits were found and exploited, including the rich deposits at Elizabethtown (Colfax County) and Pinos Altos (Grant County). Many discoveries of rich lode deposits followed as a result of the placer discoveries. By the end of the 19th century, many of the placers discovered were already exhausted. In 1901 a prospector located the placers at the foot of the Caballos Mountains in Sierra County and tried (unsuccessfully) to keep the location of his rich find a secret. By 1903 the secret was out, and a gold rush to the Caballos Mountains followed. In 1908 the discovery of placers at Sylvanite (Hidalgo County) caused the last gold rush in New Mexico.

Development of some placer deposits, and abandonment of many others, continued until the depression years of the early 1930's. Placer mining all over the West underwent a great revival during the depression; many individuals turned to placer deposits to earn a grub stake or just a meal ticket. During this period much attention was given to the invention and development of a myriad of jigs, drywashing machines, and separation methods for recovery of gold from placers and the literature describing these techniques is voluminous. After the boom of the 1930's, the war years of the 1940's were a setback to gold mining activity. War Board Order L–208 greatly restricted the development of gold mines; prospecting and mining metals essential to the war effort was deemed more important than mining for gold. More important, however, the economy of the 1940's encouraged work in offices, factories, war industries, and supporting industries, for those not in military service. Many miners and prospectors left the field for the cities and never returned. New Mexico, never a rich gold placer State in comparison with California or Oregon, never again attained the productivity in placer mining it once knew.

PURPOSE AND SCOPE OF PRESENT STUDY

The present paper is a compilation of published information relating to the placer gold deposits of New Mexico, one of a series of four papers describing the gold placer deposits in the Southwestern States. The purpose of the paper is to outline areas of placer deposits in New Mexico and to serve as a guide to their location, extent, production history, and source. The work was undertaken as part of the investigation of the distribution of known gold occurrences in the Western United States.

Each placer is described briefly. Location is given by geographic area and township and range. Topographic maps and geologic maps which show the placer area are listed. Access to each area is indicated by direction and distance along major roads and highways from a nearby center of population.

Detailed information relating to the exact location of placer deposits, their thickness, distribution, and average gold content (all values cited in the text have been converted to gold at $35 per ounce, except where otherwise noted) is included, where available, under the section entitled "Extent."

Discovery of placer gold and subsequent placer-mining activity are briefly described in the section entitled "Production History." Detailed discussion of mining operations is omitted, as this information can be found in the individual papers published by the State of New Mexico, in the yearly Mineral Resources and the Mineral Yearbook volumes published by the U.S. Bureau of Mines and the U.S. Geological Survey, and in many mining journals. Placer gold production, in ounces (table 1) was compiled from the yearly Mineral Resources and Mineral Yearbook volumes and from information supplied by the U.S. Bureau of Mines. These totals of recorded production are probably lower than actual gold production, because substantial amounts of coarse placer gold commonly sold by indivduals to jewelers and specimen buyers are not reported to the U.S. Bureau of Mines or to the U.S. Bureau of Mint. Information about the age and type of lode deposit that was the source of the placer gold is discussed for each district.

A detailed search of the geologic and mining literature was made for information concerning all the placers. A list of literature references is given with each district; the annotation indicates the type of information found in each reference. Sources of information are detailed reports on mining districts, general geologic reports, Federal and State publications, and brief articles and news notes in mining journals. There are five excellent general source books for the mining districts of New Mexico—Jones, 1904; Lindgren, Graton, and Gordon, 1910; Lasky and Wooten, 1933; Anderson, 1957; and Howard, 1967. These source books give the location and history of most placer districts in the State; many other publications give additional information about the placers. A complete bibliography, at the end of the paper, includes separate sections for all literature references and all geologic map references.

Map publications of the Geological Survey can be ordered from the U.S. Geological Survey, Distribution Section, Denver Federal Center, Denver, Colo. 80225; book publications, from the Superintendent of Documents, Government Printing Office, Washington, D.C. 20402.

COLFAX COUNTY

1. ELIZABETHTOWN DISTRICT

Location: Moreno River Valley, west flank of Baldy Mountain, Tps. 27 and 28 N., R. 16 E. (projected; on Maxwell Land Grant).

Topographic maps: Eagle Nest and Red River Pass 7½-minute quadrangles.

Geologic maps:

Bachman and Dane, 1962, Preliminary geologic map of the northeastern part of New Mexico, scale 1:380,160.

Ray and Smith, 1941, Geologic map and structure sections of Moreno Valley (pl. 1), scale 1⅛ in. = 2 miles; Physiographic map and profiles of Moreno Valley (pl. 2).

Access: From Taos, 30 miles northeast on U.S. Highway 64 to Eagle Nest. State Highway 38 leads north 5 miles to Elizabethtown and surrounding placers.

Extent: Placers are found on the slopes of Baldy Mountain, in gulches tributary to the Moreno River from the east, and in the gravels of the Moreno River. Most of the placer mining was concentrated in the area along the lower slopes and along the Moreno River Valley between Anniseta Gulch (2 miles south of Elizabethtown) north to Mills Gulch (3 miles north of Elizabethtown). Some gold was recovered from gulches on the west side of the Moreno River before 1900 (West Moreno; Hematite district).

The sediments in the Moreno River Valley consist of a thick sequence (more than 300 ft. thick) of locally derived unconsolidated sand and gravels that range in age from Pliocene(?) to Holocene. Although the geology of the placer gravels has not been studied in detail, some of these deposits, such as those exposed in deep pits about a quarter of a mile east of Elizabethtown, are believed to be correlative with the Eagle Nest Formation (Pliocene) exposed in the southern part of the Moreno River Valley (Ray and Smith, 1941). The Eagle Nest Formation is considered to be a series of coalescing stream fan deposits which filled the valley during the late Tertiary. The placer gravels on the mountain slopes were generally only a few feet thick and were confined to narrow gulches. Gold, in both the deep river gravels and the shallow slope gravels, was concentrated on the surfaces of hard clay layers, in rich lenses in gravel layers, and in crevices in decomposed bedrock.

Production history: Placer gold valued at more than $3 million was recovered from this district. The Moreno River, Grouse and Humbug Gulches, and Spanish Bar (opposite the mouth of Grouse Gulch) were the most productive placer areas in the district. The greatest part of the gold was recovered during the period 1866–1904. Most of the gold was mined by small-scale sluicing and hydraulic methods, but dredge operations along the Moreno River during the period 1901–5 recovered the major part of gold produced in New Mexico during that period. During this century, most of the placer mining has been on a small scale.

Source: The placer gold in the Elizabethtown district was derived from numerous gold-bearing veinlets which occur in the porphyry bedrock on the west slope of Mount Baldy. This bedrock area has supplied detritus to the Moreno River Valley since the late Tertiary. The veinlets occur throughout the porphyry but in only a few places, as at the Red Bandana group of mines, are they large enough to make lode mining profitable. The quartz-pyrite fissure veins which constitute the Red Bandana lodes are thought to be the principal source of gold in the Grouse Gulch gravels.

Literature:

Anderson, 1956: Production estimates.

——— 1957: General history; production.

Burchard, 1882: Production estimates.

Frost, 1905: Production of El Oro Dredge on Moreno River.

Graton, 1910: Describes extent, value, and origin of placers.

Howard, 1967: Production information for the placers.

Jones, 1903: Describes dredge operations on Moreno River.

——— 1904: Describes history and production; describes placer claims.

Koschmann and Bergendahl, 1968: Production information.

Lasky and Wooton, 1933: Describes thickness of gold-bearing gravels.

Metzger, 1938: Describes problems related to placer mining.

Pettit, 1966a: Ownership history.

——— 1966b: Mining history.

Ray and Smith, 1941: Describes geology of gravels.

Raymond, 1870: Detailed descriptions of placer claims in Grouse and Humbug Gulches.

——— 1872: Production information for 1870.

——— 1873a: Production information for 1871.

——— 1873b: Production information for 1872.

——— 1877: Production information for 1875.

Wells and Wooton, 1932: Extent of gold-bearing gravels.

2. MOUNT BALDY PLACERS

[The Mount Baldy placers here include placers found in the major streams draining the flanks of Baldy Mountain—Willow, Ute, and South Ponil Creeks. Many writers separate the area into small districts named after these major streams]

Location: West, south, and east flanks of Baldy Mountain, T. 27 N., Rs. 16–18 E. (projected; on Maxwell Land Grant).

Topographic map: Ute Park 15-minute quadrangle.

Geologic maps:

Bachman and Dane, 1962, Preliminary geologic map of the northeastern part of New Mexico, scale 1:380,160.

Wanek, Read, Robinson, Hays, and McCallum, 1964, Geologic map of the Philmont Ranch region, New Mexico, scale 1:48,000. (See also Robinson and others, 1964.)

Access: From Taos, 42 miles northeast on U.S. Highway 64 to Ute Park and vicinity. Light-duty and dirt roads lead from U.S. Highway 64 to the placer areas in Willow, Ute, and South Ponil Creeks.

Extent: The placers are found in the gravels of Willow, Ute, and South Ponil Creeks. Willow Creek drains the southwest flank of Baldy Mountain; the upper part of the stream flows through a narrow valley, the lower part over a large alluvial fan built up by debris eroded from the mountains by the creek. Most of the placer mining along Willow Creek was concentrated along the upper reaches of the stream, where gold was found in the creek and hillside gravels. Ute Creek drains the southeast flank of Baldy Mountain; placer mining was concentrated in the creek gravels between the Aztec mine downstream to the Atmore Ranch. South Ponil Creek drains the east flank of Baldy Mountain; the placers in this creek were apparently found in the upper reaches near the outcrop of the Aztec vein. Small placers were worked in North Ponil Creek, about 10 miles east of Mount Baldy.

Production history: Placer gold valued at more than $1 million was produced from this area. Most of the production was obtained from Willow and Ute Creeks. Most of the placer mining was on a small scale, although dredges worked on both Willow and Ute Creeks for a few years.

Source: The gold in Willow Creek is derived, at least in part, from small gold-bearing veinlets found in the porphyry bedrock near the northwest fork of Willow Creek. The gold in Ute and South Ponil Creeks is believed to be derived from the eroded outcrops of the Aztec vein, which is located on the ridge separating the two creeks.

Literature:

Anderson, 1956: Production estimates.

———— 1957: Brief outline of placer mining operations; production.

Burchard, 1882: Production estimates.

Chase and Muir, 1923: Discovery of placers on Ute Creek.

Howard, 1967: Production information for placers.

Jones, 1904: Describes history and production; describes gold in Ute Creek placers.

Koschmann and Bergendahl, 1968: Production information.

Lasky and Wooton, 1933: Source of gold in Ponil placers.

Lee, 1916: Placer discovery.

Metzger, 1938: Describes placer mining in 1935.

Pettit, 1966a: Ownership history.

———— 1966b: Mining history.

Raymond, 1870: Details of placer claims in Willow Creek.

———— 1872: Production information for Willow Creek placers for 1870.

Robinson, Wanek, Hays, and McCallum, 1964: Brief description of Ute Creek placers.

3. CIMARRONCITO DISTRICT

Location: South flank of Black Mountain in the Cimarron Range, south of the Cimarron River, T. 26 N., R. 18 E. (projected; on Maxwell Land Grant).

Topographic map: Tooth of Time 15-minute quadrangle.

Geologic maps:

Bachman and Dane, 1962, Preliminary geologic map of the northeastern part of New Mexico, scale 1:380,160.

Wanek, Read, Robinson, Hays, and McCallum, 1964, Geologic map of the Philmont Ranch region, New Mexico, scale 1:48,000.

Access: From Taos, 57 miles northeast on U.S. Highway 64 to Cimarron and junction with State Highway 21; from there, 3 miles south to Philmont Scout Ranch Headquarters. Dirt roads lead to vicinity of Black Mountain and Urraca Creek.

Extent: Placers were reported on Urraca Creek and tributaries.

Production history: Placers were apparently worked in 1898, but no production was recorded.

Source: Unknown.

Literature:

Anderson, 1957: Lists Cimarroncito as placer district.

Jones, 1904: Locates gold-bearing creeks.

Mining Reporter, 1898a: Placer mining in 1898.

GRANT COUNTY

4. WHITE SIGNAL DISTRICT

[White Signal district here includes placer deposits of districts previously named Malone (Gillerman, 1964; Anderson, 1957; Jones, 1904) or southwestern and central Big Burro Mountains (Gillerman, 1964); the deposits are included in the White Signal district by Howard (1967)]

Location: South flank of the Big Burro Mountains, T. 20 S., Rs. 14–16 W.

Topographic maps: Burro Peak and White Signal 7½-minute quadrangles; Redrock 15-minute quadrangle.

Geologic maps:

Ballman, 1960, Geology of the Knight Peak area, scale 1:63,360.

Dane and Bachman, 1961, Preliminary geologic map of the southwestern part of New Mexico, scale 1:380,160.

Gillerman, 1964, Geologic map of western Grant County (pl. 1), scale 1 : 126,720.

Access: From Lordsburg, 29 miles northeast on State Highway 180 to White Signal. Dirt roads lead from main highway at various points to vicinity of different placer localities.

Extent: Placers in the White Signal district are found in two areas: (1) along Gold Gulch and Thompson's Canyon (T. 20 S., R. 16 W., Redrock 15-minute quadrangle; Burro Peak quadrangle) especially in that part of Gold Gulch which traverses secs. 21 and 22 (T. 20 S., R. 16 W., Burro Peak quadrangle; Sunset Goldfields placer) and in a small tributary to Gold Gulch (sec. 27, T. 20 S., R. 16 W., Burro Peak quadrangle; Cureton placer); and (2) in the vicinity of Gold Lake (sec. 20, T. 20 S., R. 14 W., White Signal quadrangle) about 10 miles east of Gold Gulch.

There have been reports of placer gold found in unspecified streams and drywashes within the Burro Mountains. One report indicates that 2 ounces of placer gold was recovered from the Paymaster claim (attributed to the Burro Mountain district in 1942), which may be part of the Paymaster group of lode claims (secs. 21 and 28, T. 20 S., R. 15 W., Burro Peak quadrangle).

Production history: The placers in Thompson's Canyon and Gold Gulch have been worked at least since 1884, and early production is unknown. Most of the activity in this century occurred during the 1930's under the direction of a small placer mining company. The Gold Lake Placer was worked during the period 1900–10 and again during the period 1931–32.

Source: The origin of the gold in Thompson's Canyon and Gold Gulch is unknown, but probably is in gold veins that occur in the adjacent mountains. The gold found in the Gold Lake area was derived from small veinlets in a small knob of granite which protrudes through the alluvium at the lake.

Literature:

Anderson, 1957: Reports placer occurrence.

Ballman, 1960: Describes geology of Gold Gulch area.

Burchard, 1885: Reports that placers in Thompson's Canyon and Gold Gulch produced large amounts of gold.

Gillerman, 1964: Describes placers in White Signal district.

Howard, 1967: Locates placers.

Jones, 1904: Describes placer occurrence.

Raymond, 1877: Reports placers in the Burro Mountains.

U.S. Bureau of Mines, 1941: Reports production of placer gold in the Burro Mountains.

———— 1942: Reports production of placer gold in the Burro Mountains. .

5. PINOS ALTOS DISTRICT

Location: Pinos Altos Mountains, Tps. 16 and 17 S., Rs. 13 and 14 W.

Topographic maps: All 7½-minute quadrangles—Fort Bayard, Reading Mountain, Twin Sisters, Silver City.

Geologic maps:

Dane and Bachman, 1961, Preliminary geologic map of the southwestern part of New Mexico, scale 1 : 380,160.

Paige, 1911, Geologic relations of fissure veins near Pinos Altos (fig. 10), scale approximately 1 in. = 1 mile.

Access: From Silver City, 8 miles north-northeast on State Highway 25 to the town of Pinos Altos. Dirt roads lead from the town into surrounding hills and placer localities.

Extent: The placers in the Pinos Altos district are found in the vicinity of the sulfide-gold-silver veins in the district, and the gold is commonly concentrated in the gulches below the oxidized vein outcrops. The richest placers were found in Bear Creek (especially near sec. 30, T. 16 S., R. 13 W., Twin Sisters quadrangle), Rich Gulch (near the Mountain Key Mine, in sec. 6, T. 17 S., R. 13 W.; sec. 31, T. 16 S., R. 13 W., Fort Bayard quadrangle), Whiskey Gulch (or Rio de Arenas; there is an uncertainty regarding the position of this gulch on the Fort Bayard quadrangle) and Santo Domingo Gulch (unlocated). Many small gulches which drain near the oxidized vein outcrops were also worked for placer gold.

Production history: The placers were discovered in 1860, and they have been worked practically every year since, mostly by individuals using pans, rockers, and small sluices. Although the richest parts of the placers were probably worked out in the first few years after discovery, many miners continued to work the small gulches during the rainy seasons. Bear Creek and Santo Domingo Gulch were dredged in 1935 and during the period 1939–42, but the location of these operations is not known to me. Production from Santo Domingo Gulch in 1935 was credited to the Central district, but this was apparently an error; I have therefore changed the production table to include all production from Santo Domingo Gulch with Pinos Altos district.

Source: The gold was derived from the eroded outcrops of oxidized sulfide-gold-silver veins in the Pinos Altos district.

Literature:

Anderson, 1957: Locates placers.

Burchard, 1882: History; production.

————— 1883: Describes extent of placer mining.

Bush, 1915: Early history of placer mining.

Graton, Lindgren, and Hill, 1910: Locates placers; mining during the period 1904–5.

Hernon, 1953: Brief history.

Howard, 1967: History; placer mining during the period 1939–41.

Jones, 1904: Describes history and methods of placer mining.

Koschmann and Bergendahl, 1968: Production estimates.

Lasky and Wooton, 1933: Production estimates.

Paige, 1911: Sketch map on page 110 locates the principal lode mines near which the placers were found.

———— 1916: Geology of recent gravels.

Raymond, 1870: Early history of discovery and production of placers.

———— 1872: Production information for 1870.

Schilling, 1959: Road log locates placers.

U.S. Bureau of Mines, 1935: Reports dredge operations on Santo Domingo Creek.

———— 1941: Reports dredge operations on Santo Domingo Creek.

———— 1942: Reports dredge operations on Santo Domingo Creek.

Wolle, 1957: Placer gold recovered in 1955.

Wright, 1915: History.

Wells and Wooton, 1932: Reports black sand analyses.

6. BAYARD AREA

Location: Southeast of the Pinos Altos Mountains, Tps. 17 and 18 S., Rs. 12 and 13 W.

Topographic maps: Fort Bayard and Santa Rita 7½-minute quadrangles.

Geologic maps:

Dane and Bachman, 1961, Preliminary geologic map of the southwestern part of New Mexico, scale 1:380,160.

Lasky, 1936, Geologic map of the Bayard area and outcrops of veins and faults in the Bayard area (pls. 1, 9), scale 1:12,000.

Access: From Silver City, 10 miles east on U.S. Highway 260 to Bayard. Dirt roads lead to mining areas and small placers.

Extent: Placers are found in almost every arroyo that drains the mineralized area in the vicinity of Bayard. The gold is concentrated in workable quantities in only a few areas: (1) south of the Copper Glance vein (secs. 32 and 33, T. 17 S., R. 12 W.; sec. 5, T. 18 S., R. 12 W., Santa Rita quadrangle); (2) the downslope side of the Owl-Dutch Uncle-Tin Box-Lost Mine vein linkage, especially along Gold Gulch (secs. 31 and 32, T. 17 S., R. 12 W.; sec. 5, T. 18 S., R. 12 W.; sec. 6, T. 18 S., R. 12 W., Santa Rita quadrangle), and (3) the vicinity of the veins along Highway 180 between Bayard and Central (sec. 1, T. 18 S., R. 13 W., Fort Bayard quadrangle).

Production history: The Central district is the most productive mining region in New Mexico and includes the areas or subdistricts of Bayard, Santa Rita, Georgetown, and Hanover, where copper, lead, and zinc are mined. The Bayard area is the only part of the Central district with a reported production of placer gold. The small amount of gold credited to the area was recovered by individuals working intermittently over a period of many years.

Source: The gold was derived from the oxidized parts of the quartz-sulfide veins in the district.

Literature:

Frost, 1905: Reports placer mining on Whitewater Creek.

Howard, 1967: Describes source of placer gold.

Lasky, 1936: Describes placers in Bayard area.

HIDALGO COUNTY

7. SYLVANITE SUBDISTRICT

Location: West flank of the south half of the Little Hatchet Mountains, T. 28 S., R. 16 W.

Geologic maps:

Dane and Bachman, 1961, Preliminary geologic map of the southwestern part of New Mexico, scale 1:380,160.

Lasky, 1947, Geologic and topographic map of the Little Hatchet Mountains (pl. 1), scale 1:31,250.

Access: From Lordsburg, 20 miles southeast on U.S. Highway 70–80 to the junction with State Highway 81; from there about 20 miles south to Hachita. Placer ground lies south of Highway 3 (9).

Extent: Placers are found in shallow draws and gulches in gravel remnants on monzonite bedrock between Cottonwood Spring and Livermore Spring (secs. 21 and 28, T. 28 S., R. 16 W.).

Production history: The placers in the Sylvanite subdistrict were discovered in 1908. The discovery created enough excitement to cause a gold rush to the area, but the placers were rapidly worked out. Most production was in 1908, but some placer mining was done in this century at the Bader placer (SW. cor. sec. 21, T. 28 S., R. 16 W.). One report notes that placer gold was discovered at Hachita in 1880.

Source: The gold was derived from the eroded outcrops of the telluride-native gold veins.

Literature:

Anderson, 1957: History.

Dinsmore, 1908: History of discovery.

File and Northrup, 1966; Placer gold at Hachita, 1880.

Hill, 1910: Discovery; extent; production.

Jones, 1908a: Discovery; extent; size of nuggets; production.

——— 1908b: Virtually repeats his article in Engineering and Mining Journal, v. 86, 1908.

Lasky, 1947: Location; extent; geology; age and origin of placer gravels; source of gold; fineness and size of gold particles; placer-mining operations.

LINCOLN COUNTY

8. JICARILLA DISTRICT

Location: Jicarilla Mountains, T. 5 S., R. 12 E.

Topographic map: Roswell 2-degree sheet, Army Map Service.

Geologic maps:

Dane and Bachman, 1958, Preliminary geologic map of the southeastern part of New Mexico, scale 1:380,160.

Griswold, 1959, Geologic map of Lincoln County (pl. 2), scale 1 in. =6 miles; Generalized geologic map of Jicarilla district (fig. 21), scale approximately 1 in. = 1 mile.

Access: Three miles north of Carrizozo, a light-duty road leads east from U.S. Highway 54 through White Oaks Canyon and through the Jicarilla Mountains.

Extent: Placers are found in gulches in the vicinity of the village of Jicarilla. Most of the placers are concentrated in Ancho, Warner, Spring, and Rico Gulches. The area is not mapped on a scale larger than 1:250,000, and the placer gulches cannot be accurately located on that scale.

Production history: The placers in the Jicarilla Mountains, discovered in 1850, have been worked on a small scale, by individuals, for more than 100 years. Although large-scale operations have not been successful because of scarcity of water and depth of overburden, mining has been profitable to the men who worked them on a small scale. From many accounts, it seems that the major part of the placer ground is unworked; however, no detailed studies of the placers have been made, and the extent of any placer ground actually remaining is unknown.

Source: The gold in the gravels is derived from small gold-pyrite veins within the monzonite porphyry intrusion, which forms the Jicarilla Mountains. In places, the gold is found directly above decomposed gold-bearing bedrock.

Literature:

Anderson, 1957: Describes extent of placers; names placer gulches; fineness of gold.

Burchard, 1883: History.

File, 1965: Lists "Rico lease" placer active in 1965.

Graton, 1910a: History; describes thickness of pay gravel and overburden in Ancho Creek placer-mining operations.

Griswold, 1959; Location; extent; names placer gulches; production; source; accessory minerals.

Jones, 1904: Placer-mining techniques; placer-mining operations.

Lasky and Wooton, 1933: Production estimates.

Raymond, 1870: Placer-mining techniques.

Smith and Dominian, 1904: States Spaniards mined Jicarilla placers (about 1700).

Wells and Wooton, 1932: Geology of placers; source; production estimates.

Wright, 1932: Gold values in gravels estimated.

9. WHITE OAKS DISTRICT

Location: In the vicinity of Baxter and Lone Mountains on the west flank of the Jicarilla Range, T. 6 S., R. 11 E.

Topographic maps: Roswell 2-degree sheet, Army Map Service; Little Black Peak and Carrizozo 15-minute quadrangles.

Geologic maps:

Dane and Bachman, 1958, Preliminary geologic map of the southeastern part of New Mexico, scale 1 : 380,160.

Griswold, 1959, Geologic map of Lincoln County (pl. 2), scale 1 in. = 6 miles; Geologic map of a part of the Lone Mountain area (fig. 2).

Smith and Budding, 1959, Little Black Peak, east half, scale 1 : 62,500.

Access: 3 miles north of Carrizozo, a light-duty road leads northeast from U.S. Highway 54 through White Oaks Canyon about 8 miles to placer area.

Extent: Small placers occur in Baxter Gulch and White Oaks Gulch (secs. 35 and 36, T. 6 S., R. 11 E., projected). Placers were also found in small tributaries to these gulches in the vicinity of the lode mines in the district.

Production history: The White Oaks district has been predominantly a lode mining district; the placers were important only to individuals before the discovery of the lode in 1879.

Source: The placer gold was evidently derived from the gold-bearing quartz-pyrite veins of the district, as much of the placer mining was conducted in the vicinity of the major lodes.

Literature:

Anderson, 1957: History; location; placer-mining operations.

Graton, 1910a: History.

Griswold, 1959: Location; placer-mining problems.

Jones, 1904: History.

Smith and Dominian, 1904: Discovery of placers; includes photographs of placer area.

10. NOGAL DISTRICT

Location: Eastern side of the Sierra Blanca Mountains, southwest of Nogal, T. 9 S., Rs. 12 and 13 E.

Topographic map: Capitan 15-minute quadrangle.

Geologic maps:
   Dane and Bachman, 1958, Preliminary geologic map of the southeastern
      part of New Mexico, scale 1:380,160.
   Griswold, 1959, Geologic map of Lincoln County (pl. 2).
Access: Eight miles east of Carrizozo on U.S. Highway 380, State Highway
   37 leads 4 miles southeast to Nogal. Dirt roads lead from Nogal west to
   the placer area.
Extent: Placers are found in Dry Gulch, which drains northeast from the
   Sierra Blanca towards Nogal. The gold-bearing gravels are found about
   1 mile below the outcrops of the ore veins (probably sec. 7, T. 9 S.,
   R. 13 E.). Placers are also found at the Dugan-Dixon claim (unlocated).
Production history: The placers have been worked since 1865, but early
   production is unknown and probably is small. Minor amounts of gold
   have been recovered by sluicing along Dry Gulch during this century.
Source: The placers are found below the outcrops of the gold-sulfide fissure
   veins, mined at the Helen Rae and American lodes, and were probably
   derived from these veins or similar small gold-bearing veins.
Literature:
   Anderson, 1957: Names placer claims.
   Burchard, 1885: Production information for 1884.
   Graton, 1910a: Notes discovery of placers.
   Griswold, 1959: Placer-mining history; placer-mining operations (1957–
      58).
   Jones, 1904: Notes placer discovery.
   Raymond, 1870: Production information for 1869.

MORA COUNTY

11. MORA RIVER PLACERS (RIO LA CASA DISTRICT)

Location: Mora River Valley, in the western part of Mora County, T. 21 N.,
   R. 15 E. (projected; on the Mora Grant).
Topographic map: Santa Fe 2-degree sheet, Army Map Service.
Geologic map: Bachman and Dane, 1962, Preliminary geologic map of the
   northeastern part of New Mexico, scale 1:380,160.
Access: From Taos, about 40 miles south and southeast to Cleveland on State
   Highway 3.
Extent: Small amounts of placer gold have been found in the mountain
   gulches and in old terraces along the Mora River, near the village of
   Cleveland.
Production history: The gravels in terraces along the Mora River were
   placered before 1940, but the amount of gold recovered was small. No
   production has been recorded from this area, and all available information
   indicates that the Mora River placers are low in tonnage and gold
   content.

Source: The gold is thought to be derived from numerous quartz lenses and veinlets found near the headwaters of the Rio La Casa and Lujon Creek, 9 miles west of Mora.

Literature:

Anderson, 1957: Location; history; origin.

Harley, 1940: Location; placer-mining operations; size and fineness of gold; source.

Howard, 1967: Locates placers.

Raymond, 1870: Notes presence of placers.

OTERO COUNTY

12. OROGRANDE (JARILLA) DISTRICT

Location: Jarilla Mountains in the Tularosa Valley southeast of the White Sands National Monument, T. 22 S., R. 8 E.

Topographic maps: Orogrande North and Elephant Mountain 7½-minute quadrangles; Orogrande 15-minute quadrangle.

Geologic map: Schmidt and Craddock, 1964, Geologic map of the Jarilla Mountains (pl. 1), scale 1:24,000.

Access: From El Paso, Tex., 45 miles north on U.S. Highway 54 to Orogrande; from Las Cruces, 17 miles northeast on State Highway 3; from there 34 miles east on light-duty road to Orogrande. Dirt roads lead from Orogrande 1½ miles north to the placer area.

Extent: Placer mining was concentrated in the gravels in the NW¼ sec. 14 and NE¼ sec. 15, T. 22 S., R. 8 E. (Orogrande North quadrangle), at the south flank of Jarilla Mountains. Similar environments favorable for the concentration of placer gold occur in secs. 2, 3, 5, 9, 11, 14, and 15, T. 22 S., R. 8 E.

Production history: The Orogrande placers were actively worked in the early part of this century and again during the 1930's. Most of the work was done by individuals who dug many small holes and tunnels in the caliche cemented gravel to follow pay streaks. Most of the gold at the Little Joe claim (NE¼, NW¼ sec. 14, T. 22 S., R. 8 E.) was found in the 6–9 in. of gravel overlying bedrock.

Source: The gold in these placers was derived from complex sulfide ores formed during the closing stages of consolidation of the Tertiary monzonite adamellite, the principal rock of the district.

Literature:

Anderson, 1957: Production estimates; characteristics and values of gold in the gravels.

Gifford, 1899: Placer mining in 1899; extent of placers; size of gold particles; grade of gravels (probably exaggerated).

Graton, 1910b: Location; origin.

Howard, 1967: Origin of the placers.

Jones, 1904: Geology; placer-mining operations; size and value of gold gravels.

Lasky and Wooton, 1933: Production estimates; fineness of gold; gravel values.

Schmidt and Craddock, 1964: Detailed description of placer deposits; includes suggestions for future prospecting; extent; value of gravels; describes gravels in placer pit; distribution of gold in the gravels.

Wells and Wooton, 1932: Black sand analyses.

RIO ARRIBA COUNTY

13. HOPEWELL DISTRICT

Location: Tusas Mountains, in the valley between Jawbone and Burned Mountains southwest of Hopewell Lake, T. 29 N., R. 7 E., secs. 31 and 32; T. 28 N., R. 6 E., secs. 1 and 12.

Topographic maps: Burned Mountain 7½-minute quadrangle; Cebolla 15-minute quadrangle.

Geologic map: Bingler, 1968a, Geologic map of the Hopewell mining district (pl. 4), scale ¾ in. = 1 mile.

Access: From Santa Fe, 72 miles north on U.S. Highway 285 to Tres Piedras; from there about 15 miles east on light-duty road to Hopewell Lake and placer area.

Extent: Placers in the Hopewell district occur in gravels along Placer Creek from Hopewell Lake (sec. 32, T. 29 N., R. 7 E.) probably as far downstream as the mouth of Placer Creek (sec. 12, T. 28 N., R. 6 E.). The Fairview placer, the most extensively mined deposit, is located in the upper flat area, south of Hopewell Lake. Minor placers occur in Placer Creek Gorge and the lower flat area. Placer Creek was formerly known as Eureka Creek. Most of the area is now a campground in the Carson National Forest.

Production history: Most production occurred during the first few years after the discovery, about 1880. Gold valued at more than $175,000 was recovered during the first 3 years; total production to 1910 is estimated to be about $300,000. Mining during the 1900's has been on a small scale.

Source: The gold is derived from gold-bearing sulfide replacement veins and gold-quartz veins found in Precambrian schists and gneisses in the district. The Fairview placer formed during the early Tertiary and is now exhumed by Placer Creek. The placer gravel along Placer Creek and in the Lower Flat area is of recent alluvial origin, but some of the gold may be derived from the Fairview Placer.

Literature:

Anderson, 1957: Production history.

Benjovsky, 1945: Indicates areas with future mining potential.

Bingler, 1968: Location; extent; placer-mining operations; emphasis is on origin of placer gravels and age of deposition.

Burchard, 1884: Placer-mining operations.

Graton, 1910c: Location; history; early production; size of nuggets; depth of gravels; origin of gold.

Howard, 1967: Locates placers.

Jones, 1904: History; early placer operations; origin.

Just, 1937: Reports potential placer at the mouth of Placer Creek.

Wells and Wooton, 1932: History, production, and extent of placers; size of nuggets; depth of gravels.

14. EL RITO REGION

Location: Southwest of Valle Grande Peak in the Chama Basin, T. 25 N., R. 7 E.

Topographic maps: El Rito and Valle Grande 7½-minute quadrangles.

Geologic maps:

Bingler, 1968b, Reconnaissance geology of the El Rito 7½-minute quadrangle, scale 1:24,000.

———— 1968c, Reconnaissance geology of the Valle Grande Peak 7½-minute quadrangle, scale 1:24,000.

Access: El Rito is on State Highway 96, 12 miles north of U.S. Highway 84. The placer area is accessible by dirt road from El Rito.

Extent: Small quantities of gold occur in the conglomerate found between the drainages of El Rito Creek and Arroyo Seco.

Production history: There is no reported production from this area. The gold content of the conglomerate assays about 10¢ per ton; no commercial importance is attached to the occurrence.

Source: Unknown.

Literature:

Bingler, 1968a: Reports no evidence of prospecting or mining activity in the El Rito district.

Howard, 1967: Locates deposits.

Lasky and Wooton, 1933: Reports noncommercial concentrations of gold in the conglomerates of the Santa Fe Formation.

Wells and Wooton, 1932: Describes placer occurrence; promotion activities and assay values.

15. RIO CHAMA PLACER (ABIQUIU DISTRICT)

Location: In the valley of the Rio Chama, a few miles upstream from Abiquiu, T. 23 N., Rs. 5 and 6 E. (projected; on Juan Jose Lobato Grant).

Topographic map: Abiquiu 15-minute quadrangle.

Geologic map: Bingler, 1968a, Geologic map of Rio Arriba County, east half (pl. 1b), scale approximately ½ in. = 1 mile.

Access: The Rio Chama is accessible by dirt roads leading west from Abiquiu.

Extent: Gold reportedly occurs in the river sands and gravels along the Rio Chama a few miles above Abiquiu, but the exact location and extent of these deposits is unknown.

Production history: Placers have been worked along the Rio Chama near Abiquiu, possibly before 1848 and during the 1880's to 1900's. These deposits received some attention from mining companies before 1900, and some large-scale operations were attempted. News reports indicate that the placers were productive, although no production records are known. The district has apparently remained unworked throughout most of this century.

Source: Unknown, but thought to be derived from low-grade gold deposits in Precambrian rocks exposed along the deep canyon of the Rio Chama.

Literature:

Anderson, 1957: Location.

Bancroft, 1889: Reports knowledge of gold before 1848 near Abiquiu.

Burchard, 1885: Placer-mining activity.

Graton, 1910c: Reports placer occurrence; origin.

Howard, 1967: Reports placer occurrence.

Jones, 1904: Extent; placer-mining operations; thickness of gravels; gold values in bench gravels and river gravel.

Mining Reporter, 1898a: Placer-mining operations.

Prince, 1883: Reports placers known in 1844.

SANDOVAL, BERNALILLO, AND VALENCIA COUNTIES

[Small placer deposits occur in three widely separated localities on the flanks of the Sandia Mountains, which trend north-south through Sandoval, Bernalillo, and Valencia Counties. The presence of placers has been known for probably two centuries; many legends state that Spaniards engaged in mining in the area and probably prospected for placer gold. Despite the long history of the area, little is known of the exact location, extent, or worth of these placer deposits]

16. PLACITAS—TEJON REGION

Location: North end of the Sandia Mountains, T. 13 N., R. 5 E. (on the San Antonio de las Huertas Grant and the town of Tejon Grant).

Topographic map: San Felipe Pueblo 15-minute quadrangle.

Geologic map: Dane and Bachman, 1957, Preliminary geologic map of the northwestern part of New Mexico, scale 1:380,160.

Access: From Albuquerque, 17 miles north on U.S. Interstate 25 to State Highway 44; from there, about 7 miles east to Placitas (Sandoval County).

Extent: Placers are found in the area around the towns of Placitas and Tejon, at the north end of the Sandia Mountains in the vicinity of Las

Huertas Creek and Tejon Canyon (center sections of T. 13 N., R. 5 E.).

Production history: Gold reportedly occurs in beds of cemented gravels in this region; individuals using drywashing machines reportedly recovered gold worth $3 per day per man during the first decade of this century.

Source: Unknown.

Literature:

Anderson, 1957: Mentions presence of placers.

Elston, 1967: Describes vein mineralization; no placer information.

Heikes and York, 1913: Placer-mining operations; production; type of gravels.

Jones, 1904: Reports occurrence of auriferous cemented gravels.

Wells and Wooton, 1932: Production information for 1904.

17. TIJERAS CANYON REGION

Location: Central part of the Sandia Mountains, T. 10 N., Rs. 4–6 E.

Topographic maps: Albuquerque 15-minute quadrangle; Tijeras 7½-minute quadrangle.

Geologic map: Dane and Bachman, 1957, Preliminary geologic map of the northwestern part of New Mexico, scale 1:380,160.

Access: From Albuquerque, about 18 miles east on U.S. Highway 66 to Tijeras Canyon (Bernalillo County).

Extent: The dry streams around Tijeras Canyon and the alluvial flats between the Sandia Mountains and Albuquerque on the west have been drywashed intermittently.

Production history: No production has been recorded from this area. The area is now included in the Sandia Military base and is not accessible to prospectors.

Source: The gold was probably derived from small quartz lenses in Tijeras Canyon formed during the Precambrian mineralization which contain native gold in the oxidized parts of the veins.

Literature:

Anderson, 1957: Lists as placer district; no description.

Burchard, 1882: Reports placer excitement in Tijeras Canyon area in 1881.

———— 1884: Reports placer excitement in the Rio Grande north of Albuquerque in 1883; size of gold nuggets found.

Elston, 1967: Describes bedrock mineralization; no placer descriptions.

Howard, 1967: Placer prospecting history.

18. HELL CANYON REGION

Location: Southern part of the Sandia Mountains, T. 8 N., Rs· 3–5 E. (on the Isleta Pueblo Grant).

Topographic map: Mount Washington 7½-minute quadrangle.

Geologic maps:
  Dane and Bachman, 1957, Preliminary geologic map of the northwestern part of New Mexico, scale 1:380,160.
  Reiche, 1949, Geologic map of the Manzanita and North Manzano Mountains (pl. 5), scale approximately 1 in. = 1 mile.
Access: From Albuquerque about 13 miles south on U.S. Interstate 25 to Isleta; from there dirt roads lead east along Hell Canyon (Valencia County).
Extent: Placers are found in the gravels in Hell Canyon and other drywashes in that vicinity. A narrow strip of placer ground is located at the west end of the Milagras group of patented claims (sec. 29, T. 8 N., R. 5 E.).
Production history: No recorded placer production.
Source: The gold was probably derived from the gold sulfide ores mined at the Milagras group of lode claims.
Literature:
  Jones, 1904: Reports placer occurrence; placer-mining development.
  Reiche, 1949: Locates placer claim in Hell Canyon; describes lodes.

SAN MIGUEL COUNTY

19. WILLOW CREEK DISTRICT

Location: East side of the Pecos River in the Sangre de Cristo Range, T. 18 N., Rs. 12 and 13 E.
Topographic map: Cowles 7½-minute quadrangle.
Geologic maps:
  Bachman and Dane, 1962, Preliminary geologic map of the northeastern part of New Mexico, scale 1:380,160.
  Miller, Montgomery, and Sutherland, 1963, Geology of part of the southern Sangre de Cristo Mountains, New Mexico (pl. 1), scale 1:63,360.
Access: From Pecos about 12 miles north to junction of Willow Creek and the Pecos River.
Extent: Willow Creek is frequently listed as a placer locality, but no description of the deposits has been found in the literature. The Pecos Copper mine is located at the junction of Willow Creek and the Pecos River; the ores contain a complex mixture of sulfides, gold, and silver. It is possible that some placer gold was recovered from debris eroded from this lode.
Production history: The presence of placer gold in this area has been known for many years, but there is no reported placer production.
Source: Unknown.
Literature:
  Anderson, 1957: Lists as placer district; no description.

Burchard, 1883: Reports presence of gold in streams draining Sangre de Cristo Mountains.

Lasky and Wooton, 1933: Lists as placer district; no description.

20. VILLANUEVA AREA

Location: Northeast side of the Pecos River between Villanueva and Sena, T. 12 N., R. 15 E.

Topographic map: Villanueva 15-minute quadrangle.

Geologic map: Johnson, 1970, Geologic map of the Villanueva quadrangle, scale 1:62,500.

Access: Villanueva is located on State Highway 3, 12 miles south of U.S. Highway 85.

Extent: A fossil placer occurs within crossbedded layers in the Permian sandstone (probably Yeso Formation, but referred to as Glorieta Formation by Harley, 1940) exposed in bold cliffs along the Pecos River between Sena and Villnueva (NE¼ T. 12 N., R. 15 E.). The placers were prospected before 1940 by three tunnels driven into lenses of crossbedded sandstone between layers of thin impervious shaly sandstone; small amounts of gold are concentrated in small pockets within the lenses.

Production history: Assays made on samples from the sandstone show variable amounts of gold—as high as $22.00 per ton and as low as a trace of gold; the deposit is considered uneconomic.

Source: Unknown. The Precambrian rocks exposed north of Villanueva might contain a small amount of gold in quartz veins and might have been the source of this deposit.

Literature:

Harley, 1940: Location; extent; prospecting activity; distribution of gold in sandstone; origin; assays.

21. LAS VEGAS AREA

Location: Within the city of Las Vegas, T. 16 N., R. 16 E. (projected; within the Las Vegas Grant).

Topographic map: Las Vegas 7½-minute quadrangle.

Geologic map: Bachman and Dane, 1962, Preliminary geologic maps of the northeastern part of New Mexico, scale 1:380,160.

Access: From Santa Fe, about 55 miles south and east on U.S. Highway 85–84 to Las Vegas.

Extent: In 1883, placer gold was discovered during building excavation for the courthouse within the city of Las Vegas. A brief excitement followed, during which many placer claims were staked around the town within a radius of 2 miles.

Production history: No production has been recorded.

Source: Unknown.

Literature:

Burchard, 1884: Location; placer-mining activity; results of test samples.

SANTA FE COUNTY

22. OLD PLACERS DISTRICT (DOLORES, ORTIZ)

Location: Ortiz Mountains, Tps. 12 and 13 N., Rs. 7 and 8 E. (projected on the Ortiz Mine Grant).

Topographic map: Madrid 15-minute quadrangle.

Geologic map: Dane and Bachman, 1957, Preliminary geologic map of the northwestern part of New Mexico, scale 1:380,160.

Access: Placers are accessible by dirt roads leading east from State Highway 10 near Cerrillos (north side of mountains) or near Golden (south side of mountains).

Extent: Placers are found on the eastern and southern slopes of the Ortiz Mountains; the gold occurs in both creek and mesa gravels. The richest gravels are found on the eastern slope of the mountains in the vicinity of Dolores and Cunningham Gulch; here, the gold is in mesa gravels that are the upper part of an old debris fan formed at the mouth of Cunningham Gulch. Substantial amounts of gold were also recovered from creek gravels in Dolores Gulch and Arroyo Viejo. Placers are also found on the southern slope of the mountains, north of Arroyo Tuerto, in the vicinity of Lucas Canyon; these deposits were not so rich as those found near the town of Dolores.

Production history: The placers in the Ortiz Mountains were discovered in 1828; they have been worked on a small scale since that time. Because the gold is concentrated in pay streaks that have sporadic distribution, large-scale operations have not been successful, and the individual with the small hand-carried drywashing machine was more effective in mining the gravels than were the large companies with less mobile gold-concentrating plants. Despite the problems encountered in mining the placers, the deposits in the Ortiz Mountains have produced placer gold worth about $2 million.

Source: The placers in the district were derived from the local veins. In Cunningham Gulch, near the richest placers, two important types of ore deposits occur in brecciated margins of a trachyte-latite vent rock: gold in distinct quartz fissure veins and stringers (for example, the Ortiz mine); disseminated gold and scheelite (for example, Cunningham mine). On the south side of the mountains, contact pyrometasomatic auriferous pyrite, chalcopyrite, and scheelite are disseminated in garnet tactite (for example, Lucas and Candelaria mines—no significant production); the placers here are less rich.

Literature:

Anderson, 1957: Virtually repeats Lasky and Wooton (1933); placer mining operations after World War II.

Blake, 1859: Early mining practices.

Burchard, 1883: History; production information.

Elston, 1967: Location; repeats description by Lindgren (1910); placer mining operations during the period 1939–40 (see Smith, 1940).

Engineering and Mining Journal, 1899: Average value of gravels.

Harrington, 1939: Describes discovery; mining techniques; problems.

Hartly, 1915: Placer-mining problems at Cunningham Mesa.

Howard, 1967: Locates placer claims.

Jones, 1904: Detailed history and early production information.

Koschmann and Bergendahl, 1968: Location; extent; production.

Lasky and Wooton, 1933: Characteristics of placer gravels; gold values in black sand and concentrates.

Lindgren, 1910: Location; extent; depth of gravels.

Prince, 1883: Production from 1832 to 1835, and following.

Raymond, 1870: Location; early mining practices.

———— 1874: Raymond completed a detailed study of the Ortiz mine grant in 1873; he publishes the results of his studies on both lodes and placers in this report. The details of the extent of gold-bearing gravels and early placer-mining operations are described.

Smith, 1940: Describes large operation on placers near Dolores.

23. NEW PLACERS DISTRICT (SAN PEDRO)

Location: San Pedro Mountains, T. 12 N., R. 7 E.,

Topographic maps: San Pedro 7½-minute quadrangle; Edgewood and Madrid 15-minute quadrangles.

Geologic maps:

Atkinson, 1961, Geologic map of the San Pedro Mountains (pl. 4), scale 4 in.=1 mile.

Dane and Bachman, 1957, Preliminary geologic map of the northwestern part of New Mexico, scale 1:380,160.

Access: Placers are accessible by dirt roads leading east from State Highway 10 near Golden.

Extent: Placers are found on the north, south, and west flanks of the San Pedro Mountains; the gold is found in subangular detritus at the foot of the mountains and has been further concentrated in creeks and gulches that cut into these gravel beds. The richest gravels were found on the north side of the mountains where the gold was recovered from gravels in branches of the Arroyo Tuerto near Golden (especially in Old Timer Creek in secs. 17–20, T. 12 N., R. 7 E., Madrid quadrangle) and on the south side of the mountains where the gold was recovered from gravels in San Lazarus Creek (sec. 27, T. 12 N., R. 7 E., San Pedro quadrangle). Cemented gravels in the vicinity of Golden received considerable attention during the first part of this century, and many plans were made to extract the gold presumed to be contained in the gravels. All attempts to mine these gravels have failed.

Production history: The placers in the San Pedro Mountains were discovered in 1839; like the placers in the Ortiz Mountains, they have been worked on a small scale since that time. Large-scale operations have always been inhibited because of lack of water for wet-concentration methods, and the gravels are too wet for dry-concentration methods. Despite these difficulties, the New Placers district is credited with a production of placer gold valued at nearly $2 million.

Source: The placers were derived by erosion of (1) small shear zones and fissures in intrusive rocks of early Oligocene-to-early Miocene age; filled with quartz and auriferous pyrite containing free gold in the oxidized zone, and (2) small pockets of auriferous pyrite disseminated in tactite, containing free gold in the oxidized zone.

Literature:

Anderson, 1957: Virtually repeats information given in Lasky and Wooton; adds production information for 1931–52.

Atkinson, 1961: Placers were not examined in detail in the course of this study, and the description of the deposits is taken from earlier writers; adds commentary about production.

Blake, 1859: Early mining practices; describes nuggets.

Brinsmade, 1908: Placer-mining operations; characteristics of the gravels.

Burchard, 1883: History; production.

——— 1884: Placer-mining operations; average value of gravels.

——— 1885: Placer-mining operations; production.

Elston, 1967: Location; extent, production, and origin of the placers; average gold values in gravels; placer-mining operations.

File, 1965: Lists active placer claims.

Harrington, 1939: History; early placer-mining techniques.

Herrick, 1898: Origin of the placers.

Howard, 1967: Locates placer claims.

James, 1955: History; reports observations of a visit in 1939.

Jones, 1904: Outlines placer area and production.

——— 1906: Report on value of "cement gravels."

Lasky and Wooton, 1933: Location, extent, and origin of placer. Gold values in gravel.

Lindgren, 1910: History; placer-mining operations; extent; geology.

Koschmann and Bergendahl, 1968: Production information.

Prince, 1883: History; production information to 1845.

Raymond, 1870: Early placer-mining techniques.

Statz, 1912: Discusses origin of placers.

Yung and McCaffrey, 1903: Extent and value of gravels.

24. SANTA FE DISTRICT

Location: Sangre de Cristo Mountains, T. 17 N., R. 11 E.

Topographic map: Aspen Basin 7½-minute quadrangle.

Geologic map: Bachman and Dane, 1962, Preliminary geologic map of the northeastern part of New Mexico, scale 1:380,160.

Access: From Santa Fe, a dirt road leads up Santa Fe River to McClure Reservoir; from there, apparently no roads or trails exist.

Extent: Placer gold has been reported to occur in the upper reaches of the Santa Fe River.

Production history: No recorded production.

Source: Unknown.

Literature:

Anderson, 1957: Notes placer occurrence.

Burchard, 1884: Notes placer occurrence; location.

Elston, 1967: Notes placer occurrence and prospecting activity.

Lasky and Wooton, 1933: Notes placer occurrence.

SIERRA COUNTY

25. HILLSBORO DISTRICT (LAS ANIMAS PLACERS)

Location: Animas Hills, in the eastern foothills of the Black Range, Tps. 15 and 16 S., Rs. 6 and 7 W.

Topographic maps: Hillsboro 15-minute quadrangle; Skute Stone Arroyo 7½-minute quadrangle.

Geologic maps:

Dane and Bachman, 1961, Preliminary geologic map of the southwestern part of New Mexico, scale 1:380,160.

Harley, 1934, Topographic and geologic map of the Hillsboro (Las Animas) lode mining district (pl. 6), scale about 2½ in. = 4,000 ft; general map of the Hillsboro (Las Animas) placer mining district (pl. 7), scale approximately 1 in. = 1 mile.

Kuellmer, F. J., compiler, 1956, Geologic map of Hillsboro Peak, 30-minute quadrangle, scale 1:126,720.

Access: Placers are accessible by dirt roads leading north from State Highway 90, 5 miles east of Hillsboro and 13 miles west of Interstate 25.

Extent: Placers are found on the east and south flanks of the Animas Hills. The placers on the east flank occupy a larger area roughly bounded by Dutch Gulch on the north and the Rio Percha on the south (in Tps. 15 and 16 S., Rs. 6 and 7 W.). The placers on the south flank are found in Snake Gulch (secs. 3 and 4, T. 16 S., R. 7 W.) and Wicks Gulch (secs. 1 and 2, T. 16 S., R. 7 W.).

The major placers are concentrated on the east flank in area drained by Dutch, Grayback, Hunkidori, Greenhorn, Gold Run, and Little Gold

Run Gulches. These areas are shown on the Hillsboro quadrangle in secs. 25 and 36, T. 15 S., R. 6 W., and secs. 4–6 and 7–18, T. 16 S., R. 6 W. Surface samples of alluvium contain fine gold for 3 miles east of the junction of Dutch and Hunkidori Gulches (NE¼ sec. 33, T. 15 S., R. 6 W.).

Production history: The placers on the east flank of the Animas Hills produced placer gold valued at about $2,060,000 between 1877 and 1931. Between 1934 and 1937, large-scale placer-mining operations worked the ground along Gold Run Gulch; the success of these operations has credited the Hillsboro district with the largest amount of placer gold recovered during this century in New Mexico. The deposits in Snake and Wicks Gulches were mined by hand methods, mostly during the years immediately following the placer discovery in 1877. These deposits produced gold worth about $140,000, of which $90,000 was recovered from Wicks Gulch during the winter of 1877–78.

Source: The Tertiary (Oligocene?) andesites exposed in the Animas Hills contain gold-bearing fissure veins which are the principal source of lode gold in the district; erosion of parts of the andesite has concentrated the gold in the intermediate part of the alluvial fans on the east flank of the Animas Hills. Erosion of similar small veins has concentrated gold in Snake and Wicks Gulches on the south flank of the hills.

Literature:

Anderson, 1957: Placer-mining operations during the period 1935–42.

Burchard, 1882: Placer-mining problems; production.

———— 1883: Placer-mining operations at Snake Gulch.

———— 1885: Production information for 1884.

Endlich, 1883: Source of placer gold; placer mining in 1883.

Gardner and Allsman, 1938: Placer-mining operations with movable plant; depth of gravels; clay content.

Harley, 1934: Gives a detailed description of the placers in the Hillsboro district. History, extent, geology, origin, and production of placers.

Heikes and Yale, 1913: Extent and value of placers; thickness of gravels.

Howard, 1967: Virtually repeats description of Harley (1934); notes importance of placer mining in the district.

Jones, 1904: History; early production.

Koschmann and Bergendahl, 1968: Production.

Leatherbee, 1911: History; extent; production; mining operations.

26. PITTSBURG DISTRICT (SHANDON, SIERRA CABALLO)

Location: West flank of the Caballo Mountains adjacent to the Caballo Reservoir, T. 16 S., R. 4 W.; T. 14 S., R. 4 W.; T. 17 S., R. 4 W.

Topographic maps: All 7½-minute quadrangles—Caballo, Williamsburg, Garfield.

Geologic maps:

Dane and Bachman, 1961, Preliminary geologic map of the southwestern part of New Mexico, scale 1 : 380,160.

Harley, 1934, General map of the Pittsburg (Shandon) placer mining district (fig. 19), scale approximately 1 in. = 1 mile.

Kelley and Silver, 1952, Geologic map of the Caballo Mountains (fig. 2), scale approximately 1 in. = 1 mile.

Access: From Las Cruces, 56 miles north on Interstate 25 to Caballo Dam Road; from there, 1 mile east to dirt roads that lead northeast about 1 mile to the placer area.

Extent: Placers are found in the alluvial fan near the base of the escarpment of the Caballo Mountains, in a small area in Trujillo Gulch and the area drained by its tributaries, and in Apache Canyon and Union Gulch (secs. 16 and 17, 20–22, T. 16 S., R. 4 W., Caballo quadrangle). (Trujillo Gulch is named "Caballo Canyon" on the topographic map.) Small placers reportedly occur north of the Pittsburg placer area in gulches west of Palomas Gap (T. 14 S., R. 4 W., Williamsburg quadrangle), and near Derry, south of Caballo Dam (T. 17 S., R. 4 W., Garfield quadrangle). Other than general location, I have found no information relating to the small placers at Palomas Gap and Derry.

Production history: The Pittsburg placers were discovered in 1901 by Encarnacion Silva, who attempted to keep the location secret. Until 1903, he worked the placers alone, but in November 1903, he revealed the secret at Hillsboro; the news immediately started a gold rush to the area. Many men and some small placer companies worked the gravels almost continuously from 1904 to 1941.

During the period 1932–38 placer miners recovered 32.27 ounces of gold from the Caballo Mountains, and, as this production was recorded separately from that credited to the Pittsburg district, it would appear that the other small placers were worked that year.

Source: The gold in the Pittsburg placers were derived from gold-bearing quartz veins in the Precambrian granites and schists exposed in the lower part of the escarpment; the actual source of gold is believed to have been eroded as the veins now exposed do not show large concentrations of gold.

Literature:

Anderson, 1955: Location; notes placer mining at Palomas Gap.

————— 1957: Placer operations during the period 1933–42 at Pittsburg district; lists Derry as placer district.

Gordon, 1910; Extent; geology; origin.

Harley, 1934: Extent; geology; origin.

Howard, 1967: Reports placer reserves.

Jones, 1904: History (personal account).

Kelley, 1951: Reports occurrence of placer gold in the Cambrian Bliss Sandstone.

Kelley and Silver, 1952: Describes stratigraphy of the placer deposits.

Keyes, 1903: History; geology.

27. CHLORIDE DISTRICT (UPPER CUCHILLO NEGRO PLACERS)

Location: East flank of Lookout Mountain in the Black Range, along the upper part of the Cuchillo Negro, T. 10 or 11 S., R. 9 W.

Topographic map: Lookout Mountain 15-minute quadrangle.

Geologic map: Dane and Bachman, 1961, Preliminary geologic map of the southwestern part of New Mexico, scale 1:380,160.

Access: Chloride is on State Highway 52, 29 miles west of U.S. Highway 85, 9 miles north of Truth or Consequences. Dirt roads lead from the town to surrounding areas.

Extent: Placers are reported to occur on the upper Cuchillo Negro, near Chloride. Although these deposits were discovered in 1883 and were said to have been known long before that time, I have found no information that describes the occurrence in much detail. In 1883 the bars of the stream were washed and 2–7 colors per pan were recovered.

Production history: The only recorded placer production was 1.79 ounces of placer gold attributed to the Chloride district in 1932.

Source: Unknown.

Literature:

Burchard, 1884: Placer occurrence in Chloride district.

U.S. Bur. Mines, 1932–33: Lists production for Chloride.

SOCORRO COUNTY

28. ROSEDALE DISTRICT

Location: East flank of the San Mateo Mountains, T. 6 S., Rs. 5 and 6 W.

Topographic maps: Grassy Lookout and Tenmile Hill 7½-minute quadrangles.

Geologic map: Dane and Bachman, 1961, Preliminary geologic map of the southwestern part of New Mexico, scale 1:380,160.

Access: From Truth or Consequences, 40 miles north on U.S. Highway 85 to junction with State Highway 107. Rosedale is accessible by 8 miles of dirt road leading west from State Highway 107, 18 miles northwest of U.S. Highway 85.

Extent: Unknown. Small placer was probably located in gravels near Rosedale mines.

Production history: Placer gold was recovered from stream gravels by panning in 1904 and 1905. No other information about the deposit is known.

Source: The ore in the Rosedale district was valuable only for its gold content, and consists of free milling gold in quartz veins in shear zones contained in Tertiary rhyolites. Most of the ore is oxidized and, in its higher grade, is associated with manganese oxides. The placer gold was probably derived from these oxidized ores.

Literature:

File, 1965: Reports Burris placer claim in district.

Lasky, 1932: Describes lode deposits and mining history in district.

TAOS COUNTY
29. RIO GRANDE AREA

Location: Valley of the Rio Grande from Embudo north to Red River and Cabresto Creek.

Topographic map: Taos and vicinity 30-minute quadrangle.

Geologic map: Bachman and Dane, 1962, Preliminary geologic map of the northeastern part of New Mexico, scale 1 : 380,160.

Access: From Taos, different points along the Rio Grande are accessible by local roads from State Highway 3 (north of Taos) and from U.S. Highway 64 (south of Taos).

Extent: The placer deposits occur in the riverbed, flood plain, and terrace gravels along the Rio Grande, from the county line north to the mouth of the Red River, a distance of about 33 miles. Placers also occur along the valleys of Red River, Lama Canyon, Alamo Canyon, Garrapata Canyon, San Cristobal, and the Rio Hondo.

Production history: The placers along the Rio Grande were worked during Spanish colonial times (ca. 1600). Recorded production during this century is negligible, but it is probable that much of the gold recovered was not reported to the U.S. Bureau of Mines. The placers have been the subject of study by many writers and the subject of much speculation and perhaps unwarranted investment in development. Two dredging operations (1902, and 1930's) failed because of the presence of large basalt boulders in the river gravels. The river bar deposits have yielded practically all the meager gold recovered.

Source: Unknown, but probably from gold-bearing veins in the Taos Range, east of the river. Some reports suggest that the gold was derived from Precambrian quartzites.

Literature:

Anderson, 1957: Extent; geology; placer operations.

Carruth, 1910; Summarizes earlier studies; extent and value of placers.

Deane, 1896: Placer-mining operations.

Graton and Lindgren, 1910: Notes placer occurrence.

Howard, 1967: Placer operations; source.

Schilling, 1960: Extent and character of placers; size of gold; mining operations.

Silliman, 1880: Detailed account of extent and nature of these placer gravels with estimates of grade and suggestions for operation.

30. RED RIVER DISTRICT

Location: Area around the town of Red River in the Sangre de Cristo Range, Tps. 28 and 29 N., Rs. 14 and 15 E.

Topographic maps: Taos and vicinity 30-minute quadrangle; Comanche Point and Red River 7½-minute quadrangles.

Geologic maps:

Bachman and Dane, 1962, Preliminary geologic map of the northeastern part of New Mexico, scale 1:380,160.

Schilling, 1960, Geology, mines, and prospects of the Red River mining subdistrict (pl. 1), scale 1¾ in.=1 mile.

Access: From Taos, 25 miles north on State Highway 3 to Questa, and 20 miles east on State Highway 38 to Red River. Roads lead north and south of town.

Extent: Small placers have been worked for many years in streams draining the area around the town of Red River. Near Anchor, 6 miles northeast of Red River, placers have been worked in Bitter Creek and in a side canyon extending northeast from Bitter Creek (approx. secs. 16 and 17, T. 29 N., R. 15 E., projected, Comanche Point quadrangle). Near La Belle (2 miles southeast of Anchor), small placers have been worked intermittently in Gold and Spring Creeks, tributaries to Comanche Creek. Placers have been worked intermittently for many years along Placer Creek south of Red River (T. 28 N., R. 14 E., Red River quadrangle). Gold may occur in Tertiary gravels exposed south of Placer Creek and west of the Red River on Gold Hill (approx. sec. 28, T. 28 N., R. 14 E.).

Production history: Recorded production of placer gold has been small in this century. The placers in Bitter Creek were worked during 1898 and again in the 1940's but production is unknown.

Source: Gold veins are found in both the Precambrian granites and in the Tertiary (Miocene?) quartz monzonites in the Taos Range. Although some placer gold was probably derived from the Precambrian veins, it is thought that most placer gold was derived from the Miocene quartz-pyrite veins.

Literature:

Ellis, 1931: Character of placer gold.

Engineering and Mining Journal, 1899: Placer-mining operations.

Graton and Lindgren, 1910: History; location of placer mining.

Mining Reporter, 1898a: Placer-mining operations.

——— 1898b: Placer-mining operations.

Park and McKinley, 1948: Extent; geology; placer-mining operations.

Schilling, 1960: Location; history; type of placer gravels; placer-mining operations; source.

31. RIO HONDO DISTRICT

Location: At the foot of the Sangre de Cristo Range between Arroyo Hondo and Lucero Creek, Tps. 26 and 27 N., R. 13 E. (projected).

Topographic maps: Arroyo Seco and Taos 7½-minute quadrangles.

Geologic maps:

Bachman and Dane, 1962, Preliminary geologic map of the northeastern part of New Mexico, scale 1 : 380,160.

Schilling, 1960, Geology, mines and prospects of the Rio Hondo mining district (pl. 2), scale 1⅛ in.=1 mile.

Access: From Taos, light-duty roads lead north about 8 miles to Arroyo Seco (town) between Rio Lucero and Rio Hondo. Dirt roads lead up these rivers.

Extent: Little is known about the exact occurrence and extent of these placers; apparently they were small and low grade. The placer area was in shallow surface gravels of small debris fans between Lucero Creek and Arroyo Hondo.

Production history: No recorded production.

Source: Unknown; see Red River district

Literature:

File and Northrup, 1966: States that placer gold was found in 1826.

Graton and Lindgren, 1910: Notes placer occurrence.

Howard, 1967: History.

Schilling, 1960: Gold values per cubic yard.

32. PICURIS DISTRICT

Location: Southeast side of the Rio Grande, west of the Picuris Mountains.

Topographic map: Carson 7½-minute quadrangle.

Geologic map: Montgomery, 1953, Geologic map of the Picuris Range (pl. 1.), scale 1 : 48,000.

Access: Placers are probably located in the vicinity of Pilar at the junction of State Highway 96 and U.S. Highway 64, 16 miles south of Taos.

Extent: Unkown. Placer gold recovered by the La Grande Gold Mining Co. was credited to the Picuris district; the probable location of the placer is along the Rio Grande in the vicinity of Pilar (sec. 32, T. 24 N., R. 11 E.), northwest of the Picuris Mountains.

Production History: The La Grande Gold Mining Co, produced gold from stream gravels worked by sluicing in 1908. The same company apparently installed a dredge to work gravels in the same area, or at Tres Piedras—26 miles northwest in the Tusas Mountains—in 1907. (Early mining records are frequently imprecise about locations.) Three miles southwest of Pilar, at a locality called Glenwoody, the "Oro Grande" of Pennsylvania (this company and the "La Grande" Co. are almost certainly the same) had made plans in 1902 to construct a dredge to work the river gravels above Glenwoody.

Source: Unknown. Small scattered quartz veins containing gold occur in the Precambrian rocks of the Picuris Mountains and could have supplied the gold recovered from the placers. At Glenwoody, gold reportedly occurs in a quartzite exposed in the Rio Grande cliff, and the Glen-Woody

Mining and Milling Co. was formed in 1902 to mine this supposedly large low-grade deposit. Estimates of gold values as high as $1.40 to $3 per ton were made, but mill returns apparently amounted to only 40¢ per ton and lower.

Literature:

Graton and Lindgren, 1910: Mining history at Glenwoody camp; gold values in quartzite.

Howard, 1967: History of placer mining at Glenwoody (under Rio Grande Placer Region).

Schilling, 1960: History of placer mining at Glenwoody.

U.S. Geological Survey, 1907, 1908: Placer-mining activity; operations of La Grande Gold Mining Co.

UNION COUNTY

33. FOLSOM AREA

Location: Cimarron River Valley, northeast of Folsom, T. 31 N., R. 31 E.

Topographic map: Dalhart 2-degree sheet, Army Map Service.

Geologic map: Bachman and Dane, 1962, Preliminary geologic map of the northeastern part of New Mexico, scale 1 : 380,160.

Access: State Highway 325 leads 20 miles northeast from Folsom to the Cimarron River Valley and placer area.

Extent: Gravels derived from erosion of basalt lava in the Cimarron River Valley contain small amounts of placer gold. The basalt is about 20 miles northeast of Folsom and flowed within the river valley for several miles. The exact location of the gold veinlets and placers near the flow is not known.

Production history: Apparently, some gold was recovered from the gravels as it is described as small flattened grains, but no production has been recorded from the area. The placer is not commercial.

Source: Small gold veinlets in basalt.

Literature:

Anderson, 1957: Notes reports of placer gold.

Harley, 1940: Describes placer occurrences.

OTHER PLACER DEPOSITS

Various surveys of New Mexico mineral resources mention the occurrence of placers in many districts or areas not described in the present report, and frequently these citations are repeated in later publications. I have consulted the literature describing those areas where placers have been reported, but have found no description of any placers. There may be small concentrations of gold in gravel deposits near many lode mines which were worked by miners for a short time, but these occurrences probably were so minor that they received only passing attention.

Chaves County.—Schrader, Stone, and Sanford (1916, p. 214), report gold along the Rio Hondo, presumably derived from deposits in Lincoln County to the west.

Grant County.—Anderson (1957, p. 20) and Lasky and Wooton (1933, p. 39) report a "Gold Camp" in Grant County; no other mention has been found of a Gold Camp in this county. A Gold Camp does occur in the Organ Mountains in Dona Ana County, but placer deposits are not mentioned for this district.

Harding County.—Harley (1940, p. 45) reports minor gold placers in the valley of Ute Creek near Gallegos, apparently discovered in the 1930's. He states that these were probably derived from basaltic flows to the north.

Hildalgo County.—Anderson (1957, p. 20) and Lasky and Wooton (1933, p. 39) list Lordsburg as a placer locality. Lordsburg has received much attention by many writers as a famous mining camp, but no mention has been found of location of placer deposits.

Quay County.—Jones (1904, p. 19) points out that a much-publicized "gold strike" in January 1904 was a hoax. The gravels on Reveulto Creek, 18 miles east of Tucumcari, were salted.

Socorro County.—Howard (1967, p. 169) reports a Kolosa placer claim in T. 5 S., R. 6 E., in the Mound Springs district (west of the Sierra Oscura) midway between Estey City and the Jones district. No information about this placer has been found.

File (1965, p. 71) reports Placer King operations in the Silver Mountain district; the Water Canyon district (T. 3 S., R. 3 W.) is also known as the Silver Mountain district. Lasky (1932, p. 46–54) discusses the Water Canyon district and reports that sparse silver and gold mineralization is present in volcanic rocks; native gold is said to occur in a narrow vein at the Maggie Merchant claim in Shakespeare Gulch, where it is associated with galena, sphalerite, and chalcopyrite.

GOLD PRODUCTION FROM PLACER DEPOSITS

New Mexico rates ninth in the United States (seventh in the western continental States) in placer gold production. The U.S. Bureau of Mines (1967, p. 15) cites 505,000 troy ounces of placer gold produced in New Mexico from 1792–1964. However, I estimate a larger placer gold production. The largest producing districts were Elizabethtown district (Colfax County), Pinos Altos district (Grant County), Old and New Placers districts (Santa Fe County) and Hillsboro district (Sierra County). All sources of information indicate that the Elizabethtown placers have had the largest production, estimated at $5 million (Howard, 1967). During and after the depression years, the Hillsboro district was the largest producer in the State. Table 1 gives the available production information for 33 placer districts. For comparison, I have included, as table 2, a list of 17 gold districts in New Mexico that have produced more than 10,000 ounces of gold (from Koschmann and Bergendahl, 1968).

TABLE 1.—*New Mexico placer gold production, in ounces*

Map locality (pl. 1)	County and placer district	Estimated production, discovery to 1902	Recorded production (data from U.S. Bur. Mines)			Total recorded production 1902–68	Total estimated production	Reference source for estimated production
			1902–33	1934–42	1943–68			
	Colfax:							
1, 2	Elizabethtown and Mount Baldy.	225,000	23,508	1,611	48	25,167	250,000	Howard (1967).
3	Cimarroncito	Unknown	0	0	0	0	Unknown	
	Grant:							
4	White Signal	Minor	21	345	0	366	<1,000	Lasky and Wooton (1933).
5	Pinos Altos	38,842	4,122	1,771	102	5,995	50,000	
6	Bayard area	Unknown	30	79	19	128	<1,000	
	Hidalgo:							
7	Sylvanite	0	106	3	0	109	<200	
	Lincoln:							
8	Jicarilla	4,500	1,150	1,868	2	3,020	8,000	Do:
9	White Oaks	Unknown	860	25	0	885	1,000	
10	Nogal	Unknown	50	84	0	134	200	
	Mora:							
11	Mora River placers	0	0	0	0	0	0	
	Otero:							
12	Orogrande	400	972	564	10	1,546	>2,000	Lindgren, Graton, and Gordon (1910).
	Rio Arriba:							
13	Hopewell	15,000	81	40	0	121	~16,000	Do:
14	El Rito region	Unknown	0	0	0	0	0	
15	Rio Chama	Unknown	>0	0	0	>0	<100	
	Sandoval, Bernalillo, and Valencia:							
16	Placitas-Tejon	Unknown	49	0	0	49	>50	
17	Tijeras Canyon	Unknown	0	0	0	0	>50	

No.	Locality							Reference
18	Hell Canyon	Unknown	0	0	0	0	>50	
	San Miguel:							
19	Willow Creek	Unknown	0	0	0	0	100	
20	Villanueva	0	0	0	0	0	0	
21	Las Vegas	0	0	0	0	0	0	
	Santa Fe:							
22	Old Placers	100,000	193	1,348	17	1,558	>100,000	Howard (1967).
23	New Placers	96,759	2,117	555	53	2,725	>100,000	Koschmann and Bergandahl (1968).
24	Santa Fe	Unknown	0	0	0	0	0	
	Sierra:							
25	Hillsboro	104,000	2,016	13,357	186	15,559	120,000	Harley (1934).
26	Pittsburg	0	3,444	3,645	0	7,089	8,000	
27	Chloride	0	2	0	0	2	2	
	Socorro:							
28	Rosedale	0	15	0	0	15	15	
	Taos:							
29	Rio Grande area	Unknown	7	9	0	16	<1,000	
30	Red River	Unknown	100	5	0	105	<500	
31	Rio Hondo	Unknown	15	0	0	15	<500	
32	Picuris	Unknown	65	0	0	65	65	
	Union:							
33	Folsom	Unknown	0	0	0	0	0	
	Total	584,501	38,923	25,309	437	64,669	661,000	
	Undistributed	----	906	0	66	972		
	State total	584,501	39,829	25,309	503	65,641	661,000	

TABLE 2.—*Major gold districts in New Mexico*

[From Koschmann and Bergendahl (1968). Production, in ounces, to 1959]

County	District	Lode production	Placer production
Bernalillo	Tijeras Canyon	34,488	
Catron	Mogollon	362,225	
Colfax	Elizabethtown-Baldy	221,400	[1] 145,138
Dona Ana	Organ	11,435	
Grant	Central	140,000	
	Pinos Altos	104,975	[1] 42,647
	Steeple Rock	>34,050	
Hidalgo	Lordsburg	223,750	
Lincoln	White Oaks	<146,500	[2]
	Nogal	<12,850	[2]
Otero	Jarilla	<16,500	[2]
Sandoval	Cochiti	41,500	
San Miguel	Willow Creek	178,961	
Santa Fe	Old Placer	[3]	99,300
	New Placer	16,000	99,690
Sierra	Hillsboro	>[3]50,000	>106,400
Socorro	Rosedale	27,750	15

[1] See table 1 for different estimate of placer gold production.
[2] See table 1 for placer gold production; Koschmann and Bergendahl do not give placer production separately:
[3] Koschmann and Bergendahl do not give lode production separately.

Lack of water available for mining purposes has hindered production in most mining districts. Large-scale dredges have operated in only a few districts: Elizabethtown district on the Moreno River, 1901–03; Hillsboro district on Gold Run Creek, 1935–42; Pinos Altos district on Bear Creek and Santo Domingo Creek, 1939–41. Most placer mining in the State was done by individuals working with small equipment such as drywashing jigs.

SUMMARY

The ultimate source of detrital gold in placer deposits is, for the most part, gold-bearing lode deposits of various types, which in New Mexico are represented by fissure veins and disseminated ores of Precambrian and Tertiary ages. The most productive lode deposits throughout the State are the fissure veins in Tertiary intrusive rocks; in some districts, gold-bearing veins in associated contact-metamorphic rocks have also yielded appreciable quantities of gold.

Most placers in New Mexico are erosional products of gold-bearing Tertiary lodes. Lindgren, Graton, and Gordon (1910, p. 75–76) emphasize the importance of the derivation of placer deposits from lode deposits of early Tertiary age. Unfortunately, I have found no information relating to absolute age dates on any of the source rocks in New Mexico. However, geologic studies have indicated that the age of the source rocks for most of the major districts is younger than Lindgren thought and probably extends from Eocene to Miocene time. The intrusion that formed Mount Baldy, Colfax County, cut Upper Cretaceous and Paleocene sedimentary rocks and therefore is younger than Paleocene. Gold-pyrite veins cut Miocene(?) intrusive rocks which have formed minor placer deposits in Taos County, 10 miles west of Mount Baldy. Gold-bearing veins in Oligocene volcanic rocks have been eroded to form rich placer deposits in the Hillsboro district of Sierra County. The intrusive rocks in the San Pedro district, Santa Fe County, are considered to be early Oligocene to early Miocene in age; the similarity between these and other intrusive rocks in the Ortiz Mountains suggests that this porphyry belt is generally the same age.

Most of the placer deposits of the State were formed during the Quaternary in parts of alluvial fans and in drainages leading from adjacent mineralized areas. The gold contained in the gravels is characteristically angular, indicating proximity to the source. However, the placers on the Mora River, and probably the placers in the Rio Grande, bear gold that appears to have traveled a long distance.

Only a few deposits are the product of erosional cycles earlier than Quaternary. Erosion has been continuous since late Tertiary in the Elizabethtown-Baldy Mountain districts (Colfax County) and the Old Placers-New Placers districts (Santa Fe County); it is likely that Tertiary placer deposits were mined in these districts. Some of the placer deposits of the Pittsburg district (Sierra County) are found in the late Tertiary Santa Fe Group; gold is also found in the Cambrian Bliss Sandstone at the Caballo Dam. The placers of the Sandia Mountains may be reworked sediments from the Santa Fe Group. The Tertiary Santa Fe conglomerate at El Rito (Rio Arriba County) and the Permian Glorieta Sandstone at Villanueva (San Miguel County) are auriferous. An auriferous conglomerate of probable Tertiary age at Hopewell (Rio Arriba County) is overlain by gravels of recent age, making it the only known economic buried stream-channel placer in the State.

Although it is true that the most productive gold placer districts of New Mexico are adjacent to the principal gold lode districts, several gold lode districts have no associated placers. A review of the literature of important lode districts suggests that two complementary processes contribute to the deposition of placer gravels.

Oxidation of gold-bearing sulfide ores helps free the gold and facilitates its erosion and sedimentary reconcentration. Districts where free gold is known to occur in the oxidized parts of fissure veins include Pinos Altos, Bayard, White Oaks, Nogal, Jicarilla, Red River, Orogrande, and Hopewell. In the Hopewell district (Rio Arriba County) and the Baldy Mountain district (Colfax County), coarse free gold is seen in quartz-pyrite veins. The ores are found at the surface in most districts, and there are several indications that large parts of many ore bodies were eroded throughout the Quaternary.

The Mogollon district (Catron County) has the largest recorded gold production in New Mexico. No placer production has been recorded from this district. The Lordsburg district (Hidalgo County) and the Willow Creek district (San Miguel County) have been mentioned as placer districts, but no placer gold production has been recorded for either. In addition, the Central district (Grant County) and the White Oaks district (Lincoln County) have produced more than 100,000 ounces of lode gold but only small amounts of placer gold.

Most of the gold produced in the Central district was a byproduct of mining base-metal ores; the small placers in the vicinity of Bayard (p. 10) were the product of erosion of gold-bearing sulfide veins situated in a very small area. No placers have been reported associated with base-metal deposits in the surrounding region. The small placers in the White Oaks district are also found in a very small area; the district, however, is predominantly a gold district, and the deposits are close to a wide alluvial plain. It is possible that gold eroded from the veins was not concentrated but rather was distributed throughout the gravels in this plain.

The gold in the Lordsburg district has been produced primarily as a by-product of base-metal ores. The Lordsburg district is composed of two sub-districts, the Virginia and the Pyramid. In the Virginia subdistrict secondary alteration and oxidation of the ores is erratic; in the Pyramid subdistrict, Quaternary alluvium covers large parts of the area. Lack of oxidation of the ores would account for the lack of placer deposits in the Virginia subdistrict, and burial by alluvial cover or distribution of eroded gold may account for lack of placer deposits in the Pyramid subdistrict.

New Mexico has been thoroughly prospected for gold deposits since 1828; all extensively exposed placer deposits have probably been found. However, there is indication that not all these deposits have been thoroughly explored and (or) exploited. Areas of placer concentration are apparently unmined at Jicarilla, Old Placers, Orogrande, Hillsboro, Pittsburg, and Hopewell districts. Extensive testing of these unmined gravels might reveal large tonnages of suitable grade to warrant future mining.

BIBLIOGRAPHY

LITERATURE REFERENCES

Anderson, E. C., 1955, Mineral Deposits and mines in south-central New Mexico, in Guidebook of south-central New Mexico: New Mexico Geol. Soc., 6th Field Conf., p. 121–122.

———— 1956, Mining in the southern part of the Sangre de Cristo Mountains, *in* Guidebook of southeastern Sangre de Cristo Mountains, New Mexico: New Mexico Geol. Soc. 7th Field Conf., p. 139–142.

———— 1957, The metal resources of New Mexico and their economic features through 1954: New Mexico Bur. Mines and Mineral Resources Bull. 39, 183 p.
General description of mining areas of the State.

Atkinson, W. W., Jr., 1961, Geology of the San Pedro Mountains, Santa Fe County, New Mexico: New Mexico Bur. Mines Bull. 77, 50 p.

Ballman, D. L., 1960, Geology of the Knight Peak area, Grant County, New Mexico: New Mexico Bur. Mines Bull. 70, 39 p.

Bancroft, H. H., 1889, History of Arizona and New Mexico, v. 17 *of* Works of Hubert Home Bancroft: San Francisco, Calif., A. L. Bancroft & Co., 829 p.

Benjovsky, T. D., 1945, Reconnaissance survey of the Headstone (Hopewell) mining district, Rio Arriba County, New Mexico: New Mexico Bur. Mines and Mineral Resources Circ. 11, 10 p.
Mining during Mexican government (p. 340); New Mexico (p. 629–801); early history of the State with valuable information on early mining developments.

Bergendahl, M. H., 1965, Gold, *in* Mineral and water resources of New Mexico: New Mexico Bur. Mines and Mineral Resources Bull. 87, 437 p.
Brief summary of placer mining in the State.

Bingler, E. C., 1968, Geology and mineral resources of Rio Arriba County, New Mexico: New Mexico Bur. Mines and Mineral Resources Bull. 91, 158 p.

Blake. W. P., 1859, Observations of the mineral resources of the Rocky Mountain chain, near Santa Fe, and the probable extent southward of the Rocky Mountain gold field: Boston Soc. Nat. History Proc. v. 7, p. 64–70.

Brinsmade, R. B. 1908, Development of San Pedro Mountain, New Mexico: Mining World, v. 28, p. 1021–1024.

Burchard, H. C., 1882, Report of the Director of the Mint upon the statistics of the production of the precious metals in the United States (for the year 1881): Washington, 765 p. [New Mexico, p. 327–353].

———— 1883, Report of the Director of the Mint upon the statistics of the production of the precious metals in the United States (for the year 1882): Washington, 873 p. [New Mexico, p. 339–389].

———— 1884, Report of the Director of the Mint upon the production of the precious metals in the United States during the calendar year 1883: Washington, 858 p. [New Mexico, p. 562–610].

———— 1885, Report of the Director of the Mint upon the production of the precious metals in the United States during the calendar year 1884: Washington, 644 p. [New Mexico, p. 373–395]
Information and statistics related to placer mining in different districts for each year.

Bush, F. V., 1915, Mining in the Pinos Altos district of New Mexico: Mining World, v. 42, p. 165–168.

Carruth, J. A., 1910, New Mexico gold gravels [Rio Grande placers]: Mines and Minerals, v. 31, p. 117–119.

Chase, C. A., and Muir, Douglas, 1923, The Aztec mine, Baldy, New Mexico [abs.]: Mining and Metallurgy no. 190, p. 33–35.

Deane, C. A., 1896, Placer deposits in New Mexico [Rio Grande Placers]: Mining Industry and Review, v. 16 [Feb. 13, 1896], p. 371–372.

Dinsmore, C. A., 1908, The new gold camp of Sylvanite, New Mexico: Mining World, v. 29, p. 670–671.

Ellis, R. W., 1931, The Red River lode of the Moreno Glacier: New Mexico Univ. Bull., v. 4, no. 3, 26 p.

Elston, W. E., 1967, Summary of the mineral resources of Bernalillo, Sandoval and Santa Fe Counties: New Mexico Bur. Mines Bull. 81, 81 p.

Endlich, F. M., 1883, The mining regions of southern New Mexico [Hillsboro district]: Am. Naturalist, v. 17, p. 149–157.

Engineering and Mining Journal, 1899, Developments in northern New Mexico: Eng. and Mining Jour., v. 68, pt. 1, p. 393.

File, L. A., 1965, Directory of mines in New Mexico: New Mexico Bur. Mines and Mineral Resources Circ. 77, 188 p.

    Lists mines and placer deposits alphabetically (p. 1–97). Locates them by mining district and county and refers briefly to other sources.

File, Lucien, and Northrop, S. A., 1966, County, Township and Range locations of New Mexico's mining districts: New Mexico Bur. Mines and Mineral Resources Circ. 84, 66 p.

    Includes list of district names, synonyms, and older names.

Frost, Max, 1905, Mining in New Mexico [Elizabethtown district]: Mining World, v. 23, p. 6–9.

Gardner, E. D., and Allsman, P. T., 1938, Power shovel and dragline placer mining: U.S. Bur. Mines Inf. Circ. 7013, 68 p.

Gifford, A. W., 1899, Wonderful dry placers [Orogrande district]: Ores and Metals, v. 8, Oct., p. 15.

Gillerman, Elliot, 1964, Mineral deposits of Western Grant County, New Mexico: New Mexico Bur. Mines and Mineral Resources Bull. 83, 213 p.

Gordon, C. H., 1910, Sierra and central Socorro Counties, in Lindgren, Waldemar, Graton, L. C., and Gordon, C. H., The ore deposits of New Mexico: U.S. Geol. Survey Prof. Paper 68, p. 213–285.

Graton, L. C., 1910, Colfax County, in Lindgren, Waldemar, Graton, L. C., and Gordon, C. H., The ore deposits of New Mexico: U.S. Geol. Survey Prof. Paper 68, p. 91–108.

Graton, L. C., 1910a, Lincoln County, in Lindgren, Waldemar, Graton, L. C., and Gordon, C. H., The ore deposits of New Mexico: U.S. Geol. Survey Prof. Paper 68, p. 175–184.

———— 1910b, Otero County, in Lindgren, Waldemar, Graton, L. C., and Gordon, C. H., The ore deposits of New Mexico: U.S. Geol. Survey Prof. Paper 68, p. 184–190.

———— 1910c, Rio Arriba County, in Lindgren, Waldemar, Graton, L. C., and Gordon, C. H., The ore deposits of New Mexico: U.S. Geol. Survey Prof. Paper 68, p. 124–133.

Graton, L. C., and Lindgren, W., 1910, Taos County, in Lindgren, Waldemar, Graton, L. C., and Gordon, C. H., The ore deposits of New Mexico: U.S. Geol. Survey Prof. Paper 68, p. 82–91.

Graton, L. C., Lindgren, Waldemar, and Hill, J. M., 1910, Grant County, in Lindgren, Waldemar, Graton, L. C., and Gordon, C. H., The ore deposits of New Mexico: U.S. Geol. Survey Prof. Paper 68, p. 295–348.

Griswold, G. B., 1959, Mineral deposits of Lincoln County, New Mexico: New Mexico Bur. Mines Bull. 67, 117 p.

Harley, G. T., 1934, Geology and ore deposits of Sierra County, New Mexico: New Mexico Bur. Mines Bull. 10, 220 p.

———— 1940, The geology and ore deposits of northeastern New Mexico (exclusive of Colfax County): New Mexico Bur. Mines and Mineral Resources Bull. 15, 102 p.

Harrington, E. R., 1939, Gold mining in the desert [New Placer and Old Placer district]: Mines Mag., v. 29, p. 508–509, 512.

Hartly, Carney, 1915, The opportunity in placer mining [Old Placer district]: Eng. and Mining Jour., v. 99, p. 185–188.

Heikes, V. C., and York, C. G., 1913, Dry placers in Arizona, Nevada, New Mexico, and California: U.S. Geol. Survey Mineral Resources U.S. [1912], pt. 1, p. 254–263.

Hernon, R. M., 1953, Summary of smaller mining districts in the Silver City region, in Guidebook of southwestern New Mexico: New Mexico Geol. Soc., 45th Field Conf. [Oct. 15–18], p. 138–141.

Herrick, C. L., 1898, The geology of the San Pedro and Albuquerque districts: New Mexico Univ. Bull., v. 1, no. 4; New Mexico Univ. Sci. Lab. Div. 7, art. 5, p. 93–116.

Hill, J. M., 1910, Sylvanite district, in Lindgren, Waldemar, Graton, L. C., and Gordon, C. H., Ore deposits of New Mexico: U.S. Geol. Survey Prof. Paper 68, p. 338–343.

Howard, E. V., 1967, Metalliferous occurrences in New Mexico: Santa Fe, N.M., State Planning Office, 270 p.
Describes mining district of the State alphabetically.

James, Henry, 1955, Muleshoe Gold [New Placer district]: New Mexico Sun Trails, v. 8, no. 1, p. 20–21.

Jones, F. A., 1903, Report of the Director of the Mint upon the production of the precious metals in the United States during the Calendar year 1902: Washington, 391 p. (G. E. Roberts, Director of the Mint); [New Mexico, p. 168–177].

———— 1904, New Mexico mines and minerals: Santa Fe, N. Mex., New Mexican Printing Co., 349 p.
Comprehensive descriptions of mining areas including placers active at that time; includes much information on history of areas.

———— 1906, Placers of Santa Fe County, New Mexico: Mining World, v. 25, p. 425.

———— 1908a, Sylvanite, New Mexico, the new gold camp: Eng. and Mining Jour., v. 86, p. 1101–1103.

———— 1908b, The new camp of Sylvanite, New Mexico: Mining Sci., v. 58, p. 489–491.

———— 1908, Epitome of the economic geology of New Mexico: Albuquerque, New Mexico Bur. Immigration, 47 p.
An outline of important mineral deposits of the State. Chief placer-mining districts are listed on pages 12–13.

Jones, F. A., 1915, The mineral resources of New Mexico: New Mexico School of Mines, Mineral Resources Survey Bull. 1, 77 p.
Revised edition of 1908 publication. Placer districts are listed on page 22; map of chief mining districts is included.

Just, Evan, 1937, Geology and economic features of the pegmatites of Taos and Rio Arriba Counties, New Mexico: New Mexico Bur. Mines Bull. 14, 73 p.

Kelley, V. C., 1951, Oolitic iron deposits of New Mexico [Pittsburgh district]: Am. Assoc. Petroleum Geologists Bull., v. 35, p. 2199–2228.

Kelley, V. C., and Silver, Caswell, 1952, Geology of the Caballo Mountains with special reference to regional stratigraphy and structure, and to mineral resources including oil and gas: New Mexico Univ. Pubs. Geology, no. 4, 286 p.

Keyes, C. R., 1903, Geology of the Apache Canyon placers: Eng. and Mining Jour., v. 76, p. 966–967.

Koschmann, A. H., and Bergendahl, M. H., 1968, Principal gold producing districts of the United States: U.S. Geol. Survey Prof. Paper 610, p. 200–211.

Describes seventeen districts in New Mexico which produced more than 10,000 ounces of gold through 1957. Major placer districts are described.

Lasky, S. G., 1932, The ore deposits of Socorro County, New Mexico: New Mexico Bur. Mines and Mineral Resources Bull. 8, 139 p.

Lasky, S. G., 1936, Geology and ore deposits of the Bayard area, Central mining district, New Mexico: U.S. Geol. Survey Bull. 870, 144 p.

———— 1947, Geology and ore deposits of the Little Hatchet Mountains, Hidalgo and Grant Counties, New Mexico: U.S. Geol. Survey Prof. Paper 208, 101 p.

Lasky, S. G., and Wooton, T. P., 1933, The metal resources of New Mexico and their economic features: New Mexico School of Mines Bull. 7, 178 p.

General description of mining areas in the State. Predecessor to Anderson, 1957.

Leatherbee, Brigham, 1911, Mesa del Oro placer grounds [Hillsboro district]: Mining World, v. 35, p. 536.

Lee, W. T., 1916, The Aztec gold mine, Baldy, New Mexico: U.S. Geol. Survey Bull. 620, p. 325–330.

Lindgren, Waldemar, 1910, Santa Fe County, in Lindgren, Waldemar, Graton, L. C., and Gordon, C. H., Ore deposits of New Mexico: U.S. Geol. Survey Prof. Paper 68, p. 163–175.

Lindgren, Waldemar, Graton, L. C., and Gordon, C. H., 1910, The ore deposits of New Mexico: U.S. Geol. Survey Prof. Paper 68, 361 p.

Detailed discussion of geology and lode and placer mining for each county in New Mexico. [Descriptions of individual counties are written by different authors and annotated separately for each county discussed in this paper.]

Metzger, O. H., 1938, Gold mining in New Mexico: U. S. Bur. Mines Inf. Circ. 6987, 71 p.

Placer deposits and operations are discussed for major placer areas.

Mining Reporter, 1898a, New Mexico placer mining: Mining Reporter, v. 38, no. 1, p. 22–23.

———— 1898b, Bitter Creek placers [Red River district]: Mining Reporter, v. 38, no. 6, p. 22.

Northrop, S. A., 1942, Minerals of New Mexico: New Mexico Univ. Bull. 379, Geology Ser., v. 6, no. 1, 387 p.

Partial list of placer gold localities is given on pages 158–159.

Paige, Sidney, 1911, The ore deposits near Pinos Altos, New Mexico: U.S. Geol. Survey Bull. 470, p. 109–125.

———— 1916, Silver City Folio: U.S. Geol. Survey Folio 199, 19 p.

Park, C. F., Jr., and McKinley, P. F., 1948, Geology and ore deposits of Red River and Twinning districts, Taos County, New Mexico: New Mexico Bur. Mines and Mineral Resources Circ. 18, 35 p.

Pettit, R. F., Jr., 1966a, Maxwell Land Grant, in Guidebook to Taos—Raton—Spanish Peaks Country: New Mexico Geol. Soc. 17th Field Conf., p. 66–68.

———— 1966b, History of mining in Colfax County, in Guidebook to Taos—Raton—Spanish Peaks Country: New Mexico Geol. Soc. 17th Field Conf., p. 69–75.

[These two articles in "Guidebook to Taos—Raton—Spanish Peaks Country" are summarized from an open file report by R. F. Pettit, Jr., entitled "Mineral Resources of Colfax County, New Mexico." The present status of the report is that it is on Open File, State Bureau of Mines and Mineral Resources, Socorro, N. Mex.]

Prince, L. B., 1883, Historical sketches of New Mexico: New York, Leggat Bros.; Kansas City, Ramsey, Millett, and Hudson, 327 p.

History of mines and mining given on pages 241–245. Details of history up to 1847.

Ray, L. L., and Smith, J. F., Jr., 1941, Geology of the Moreno Valley, New Mexico: Geol. Soc. America Bull., v. 52, no. 2, p. 177–210.

Raymond, R. W., 1870, Statistics of mines and mining of the States and Territories west of the Rocky Mountains: Washington, 805 p.,[New Mexico, p. 381–418].

———— 1872, Statistics of mines and mining in the States and Territories west of the Rocky Mountains for the year 1870: Washington, 566 p. [New Mexico, p. 282–286].

———— 1873a, Statistics of mines and mining in the States and Territories west of the Rocky Mountains (for the year 1871): Washington, 566 p. [New Mexico, p. 337–339].

———— 1873b, Statistics of mines and mining in the States and Territories west of the Rocky Mountains for the year 1872. Washington, 550 p. [New Mexico, p. 309–311].

———— 1874, Statistics of mines and mining in the States and Territories west of the Rocky Mountains for the year 1873: Washington, 585 p. (New Mexico, p. 313–344).

———— 1877, Statistics of mines and mining in the States and the Territories west of the Rocky Mountains for the year 1875: Washington, 519 p. [New Mexico, p. 337–340].

Information and statistics related to placer mining in different districts for each year.

Reiche, Parry, 1949, Geology of the Manzanita and North Manzano Mountains, New Mexico [Hell Canyon district]: Geol. Soc. America Bull., v. 60, no. 7, p. 1183–1212.

Robinson, G. D., Wanek, A. A., Hays, W. H., and McCallum, M. E., 1964, Philmont Country—the rock and landscape of a famous New Mexico ranch [Mount Baldy district]: U.S. Geol. Survey Prof. Paper 505, 152 p.

Schilling, J. H., 1959, Silver City, Santa Rita, Hurley: New Mexico Bur. Mines and Mineral Resources Scenic Trip 5, 43 p.

———— 1960, Mineral resources of Taos County, New Mexico: New Mexico Bur. Mines and Mining Resources Bull. 71, 124 p.

Schmidt, P. G., and Craddock, Campbell, 1964, The geology of the Jarilla Mountains, Otero County, New Mexico: New Mexico Bur. Mines Bull. 82, 55 p.

Schrader, F. C., Stone, R. W., and Sanford, S., 1916, Useful minerals of the United States: U.S. Geol. Survey Bull. 624, 412 p.

Gold placer districts are listed on page 214; many of those named are not described nor mentioned by later authors.

Silliman, Benjamin, Jr., 1880, Report on the newly discovered auriferous gravels of the upper Rio Grande del Norte in the counties of Taos and Rio Arriba, New Mexico: Omaha, Nebr., Herald Pub. House, 34 p.

Smith, E. P., and Dominian, Leon, 1904, Notes on a trip to White Oaks, New Mexico: Eng. and Mining Jour., v. 77, p. 799–800.

Smith, T. E., 1940, A mobile dry placer plant [Old Placers district]: Eng. and Mining Jour., v. 141, no. 6, p. 40.

Statz, B. A., 1912, The New Placer mining district, New Mexico: Mining Sci., v. 66, p. 167.

Stone, G. H., 1899, Dry gold placers of the arid regions: Mines and Minerals, v. 19, p. 397–399.

U.S. Bureau of Mines, 1925–34, Mineral resources of the United States [annual volumes, 1924–31]: Washington, U.S. Govt. Printing Office.

———— 1933–68, Minerals Yearbook [annual volumes, 1932–67]: Washington, U.S. Govt. Printing Office.

Information relating to placers cited in text is referenced by year of pertinent volume.

———— 1967, Production potential of known gold deposits in the United States: U.S. Bur. Mines Inf. Circ. 8331, 24 p.
Lists estimates of total placer gold production in troy ounces.

U.S. Geological Survey, 1896–1900, Annual reports [17th through 21st, 1895–1900]: Washington, U.S. Govt. Printing Office.

———— 1883–1924, Mineral resources of the United States [annual volumes, 1882–1923]: Washington, U.S. Govt. Printing Office.
Information relating to placers cited in text is referenced by year of pertinent volume.

Wells, E. H., 1930, An Outline of the mineral resources of New Mexico: New Mexico Bur. Mines and Mineral Resources Circ. 1, 15 p.

Wells, E. H., and Wooton, T. P., 1932, Gold mining and gold deposits in New Mexico: New Mexico Bur. Mines and Mineral Resources Circ. 5, 24 p. [Revised by T. P. Wooton, April 1940].

Report contains general information on mining and general features of placer deposits. Individual placers are described in detail for most important placer districts. Bibliography of papers relating to lode and placer gold deposits.

Wolle, M. S., 1957, Pinos Altos, New Mexico Gold Camp: Mining World, v. 19, no. 12, p. 56–57.

Wright, I. L., 1915, The Pinos Altos district, New Mexico: Eng. and Mining Jour., v. 99, p. 133–135.

Wright, P. E., 1932, The Jicarilla mining district of New Mexico: Mining Jour. [Phoenix, Ariz.], v. 16, no. 8, p. 7.

Yung, M. B., and McCaffery, R. S., 1903, The ore deposits of the San Pedro district, New Mexico: Am. Inst. Mining Engineers Trans., v. 33, p. 350–362.

GEOLOGIC MAP REFERENCE

[References keyed by number to districts given in text]

Atkinson, W. W., Jr., 1961, Geology of the San Pedro Mountains, Santa Fe County, New Mexico: New Mexico Bur. Mines Bull. 77, 50 p., pl. 4.
No. 23.

Bachman, G. O., and Dane, C. H., 1962, Preliminary geologic map of the northeastern part of New Mexico: U.S. Geol. Survey Misc. Geol. Inv. Map I–358, scale 1:380,160.
Nos. 1, 2, 3, 11, 19, 21, 24, 29–31, 33.

Ballman, D. L., 1960, Geology of the Knight Peak area, Grant County, New Mexico: New Mexico Bur. Mines Bull. 70, 39 p., pl. 1.
No. 4.

Bingler, E. C., 1968a, Geology and mineral resources of Rio Arriba County, New Mexico: New Mexico Bur. Mines and Mineral Resources Bull. 91, 158 p., pls. 1, 4.
Nos. 13, 15.

———— 1968b, Reconnaissance geology of the El Rito 7½-minute quadrangle: New Mexico Bur. Mines and Mineral Resources Geol. Map 20.
No. 14.

———— 1968c, Reconnaissance geology of the Valle Grande Peak 7½-minute quadrangle: New Mexico Bur. Mines and Mineral Resources Geol. Map 21.
No. 14.

Dane, C. H., and Bachman, G. O., 1957, Preliminary geologic map of the northwestern part of New Mexico: U.S. Geol. Survey Misc. Geol. Inv. Map. I–224, scale 1:380,160.
Nos. 16–18, 22, 23.

——— 1958, Preliminary geologic map of the southeastern part of New Mexico: U.S. Geol. Survey Misc. Geol. Inv. Map I-256, scale 1:380,160.
Nos. 8-10.

——— 1961, Preliminary geologic map of the southwestern part of New Mexico: U.S. Geol. Survey Misc. Geol. Inv. Map I-344, scale 1:380,160.
Nos. 4, 5, 6, 7, 25-28.

Gillerman, Elliot, 1964, Mineral deposits of Western Grant County, New Mexico: New Mexico Bur. Mines and Mineral Resources Bull. 83, 213 p., pl. 1.
No. 4.

Griswold, G. B., 1959, Mineral deposits of Lincoln County, New Mexico: New Mexico Bur. Mines Bull. 67, 117 p., pl. 2; figs. 21, 2.
Nos. 8-10.

Harley, G. T., 1934, Geology and ore deposits of Sierra County, New Mexico: New Mexico Bur. Mines Bull. 10, 220 p., pls. 6, 7.
Nos. 25, 26.

Johnson, R. B., 1970, Geologic map of the Villanueva quadrangle, San Miguel County, New Mexico: U.S. Geol. Survey Geol. Quad. Map GQ-869, scale 1:62,500.
No. 20.

Kelley, V. C., and Silver, Caswell, 1952, Geology of the Caballo Mountains with special reference to regional stratigraphy and structure, and to mineral resources including oil and gas: New Mexico Univ. Pubs. Geology, no. 4, 286 p., fig. 2.
No. 26.

Kuellmer, F. J., 1956, Geologic map of Hillsboro Peak: New Mexico Inst. Mining and Technology, Geologic Map 1, 30-minute quad.
No. 25.

Lasky, S. G., 1936, Geology and ore deposits of the Bayard area, Central mining district, New Mexico: U.S. Geol. Survey Bull. 870, 144 p., pls. 1, 9.
No. 6.

——— 1947, Geology and ore deposits of the Little Hatchet Mountains, Hidalgo and Grant Counties, New Mexico: U.S. Geol. Survey Prof. Paper 208, 101 p., pl. 1; fig. 2.
No. 7.

Miller, J. P., Montgomery, Arthur, and Sutherland, P. K., 1963, Geology of part of the southern Sangre de Cristo Mountains, New Mexico: New Mexico Bur. Mines and Mineral Resources Mem. 11, 106 p., pl. 1.
No. 19.

Montgomery, Arthur, 1953, Pre-Cambrian geology of the Picuris Range, north-central New Mexico: New Mexico Bur. Mines and Mineral Resources Bull. 30, 89 p.
No. 32.

Paige, Sidney, 1911, The ore deposits near Pinos Altos, New Mexico: U.S. Geol. Survey Bull. 470, p. 109-125, fig. 10.
No. 5.

Ray, L. L., and Smith, J. F., Jr., 1941, Geology of the Moreno Valley, New Mexico: Geol. Soc. America Bull., v. 52, no. 2, p. 177-210, pls. 1, 2.
No. 1.

Reiche, Parry, 1949, Geology of the Manzanita and North Manzano Mountains, New Mexico: Geol. Soc. America Bull., v. 60, no. 7, p. 1183-1212, pl. 5.
No. 18.

Schilling, J. H., 1960, Mineral resources of Taos County, New Mexico: New Mexico Bur. Mines and Mining Resources Bull. 71, 124 p., pl. 1.
Nos. 30, 31, 32.

Schmidt, P. G., and Graddock, Campbell, 1964, The geology of the Jarilla Mountains, Otero County, New Mexico: New Mexico Bur. Mines Bull. 82, 55 p., pl. 1.
No. 12.

Smith, C. T., and Budding, A. J., 1959, Little Black Peak, east half: New Mexico Inst. Mining and Technology, Geol. Map 11.
No. 9.

Wanek, A. A., Read, C. B., Robinson, G. D., Hays, W. H., and McCallum, Malcolm, 1964, Geologic map and sections of the Philmont Ranch region, New Mexico: U.S. Geol. Survey Misc. Geol. Inv. Map I–425, scale 1:48,000 (see also Robinson and others, 1964, pls. 3, 5).
Nos. 2, 3.

PART II

NEW MEXICO SCHOOL OF MINES
Richard H. Reece, President

STATE BUREAU OF MINES AND MINERAL RESOURCES
E.C. Anderson, Director

Circular No. 5

GOLD MINING AND GOLD DEPOSITS IN NEW MEXICO

By

E.H. Wells and T.P. Wootton, April 1932

Revised by T.P. Wootton, April 1940

Reissued, October 1944; May 1946

_____

Reissued (unrevised), April 1957

Socorro, New Mexico

CONTENTS

GOLD MINING AND GOLD DEPOSITS IN NEW MEXICO

By E. H. Wells and T. P. Wootton, April 1932

Revised by T. P. Wootton, April 1940

Reissued, October 1944; May 1946

Reissued (unrevised), April 1957

INTRODUCTION

One of the outstanding effects of the current business depression has been the improvement in the status of gold as compared with the other metals. Gold deposits are attracting more interest than for many years. Proved districts are being re-examined, and an active search is being made for new deposits. The State Bureau of Mines and Mineral Resources of the New Mexico School of Mines has received many requests recently for information regarding the gold resources of New Mexico, and this circular has been prepared in response to this demand.

Most of the geology and mining history prior to 1909 which is given in these notes has been taken from "The Ore Deposits of New Mexico", Professional Paper 68 of the United States Geological Survey, by Waldemar Lingren, L.C. Graton, and C.H. Gordon. Subsequent information has been obtained from various sources. A fairly complete bibliography is appended, and the reports listed contain considerable valuable data on gold in New Mexico. The "Geologic Map of New Mexico" by N.H. Darton, issued by the United States Geological Survey, has much information for the prospector with geologic training.

The information on the deposits given in this circular necessarily is hardly more than a summary. Additional information on this subject is contained in the following bulletins of the Bureau: "The Metal Resources of New Mexico and their Economic Features", by S.G. Lasky and T.P. Wootton; "The Ore Deposits of Socorro County, New Mexico", by S.G. Lasky; "The Ore Deposits of Sierra County, New Mexico", by G.T. Harley; and "The Geology of the Organ Mountains", by K.C. Dunham.

THE PRESENT STATUS OF GOLD

From January 19, 1837, through 1932, the price of gold was $20.67 plus per ounce and in 1933 the legal coinage value was continued at that price. The average weighted price per fine ounce in 1933, as computed by the U.S. Bureau of Mines, was $25.56 and in 1934 was $34.95. Under the Gold Reserve Act of 1934 the value of gold was fixed by Presidential Proclamation on January 31, 1934, at $35.00 per fine troy ounce and has remained at that figure. As a result, some lode and placer deposits of gold which were valueless a few years ago can now be worked at a profit.

PRELIMINARY INVESTIGATIONS

Although the outlook for gold mining is much improved, those who are interested in working or financing gold properties should not be unduly optimistic. Many gold deposits in New Mexico and elsewhere are still too low grade to be worked successfully. It is possible to obtain picked samples from many worthless deposits which give assay returns far greater than the average of the material that can be mined. A vein a quarter of an inch wide may actually contain $1,000 to the ton in gold, but when diluted with the adjacent country rock that must be mined with it, the resulting ore would have a gross value of less than $8 a ton. Mining, transportation, and treatment expenses for ore of this grade in many places would exceed the metal value, and mining could be carried on only at a loss. Gold gravels which would pay handsomely if at the surface may be fatally handicapped by barren overburden that must be removed before they can be reached. Placer ground that would be workable with abundant water near at hand may be valueless because of the scarcity of water or the cost of bringing it to the deposits.

New Mexico has suffered greatly from the promotion of ill-advised mining projects. Unquestionably, many statements regarding mineral deposits in the State have been made by promoters and others which would not be substantiated by efficient investigations of the deposits. Before any gold mining program requiring a large or even moderate expenditure in advance of actual production is adopted, the deposit should be carefully examined by a reliable engineer or mining geologist and a favorable report received from him. Particularly in the case of placer deposits the examination should include thorough, systematic and accurate sampling. The examination may cost hundreds or even thousands of dollars, but if it proves the property to be non-commercial, many times its cost in useless expenditures will be saved. If favorable, it should contain information of much value to the owners regarding the character and grade of the deposit, and proper methods of mining and treatment. A favorable report does not necessarily assure profitable mining, but it should indicate a reasonable probability of successful operation.

MINING AND TREATMENT OF PLACER MATERIAL

Shallow placer material is mined in open cuts with picks and shovels, plows and horse scrapers, power scrapers, dragline excavators and power shovels; and worthless overburden, if not too thick is removed by one of these methods. The gravel is transported to the gold-saving plant mechanically or by running water. Deep pay streaks can be worked by drift mining without disturbing the overlying material but at relatively increased cost.

The simplest apparatus for recovering gold from placers is the miner's pan, and it is indispensable in prospecting. Its capacity is small - about one cubic yard per day - but in the hands of a skillful operator it is quite efficient. A surprising amount of gold has been obtained by its use.

The rocker is used in prospecting and inworking small placer deposits. The gravel is passed through the machine with a rocking motion, and the gold collects against riffles or on an apron. The rocker is easily and cheaply constructed and obtains good gold recoveries with a moderate amount of water. It gives best results when operated by two men, who can handle from three to five cubic yards of gravel per day with it.

The sluicing method of recovering gold is practiced where the deposit is of at least moderate size and water is abundant. The sluice (also called riffle box) is a slightly inclined trough with cross-pieces or other obstructions in the bottom, called riffles. The gravel is deposited in the head box at the upper end and washed through the sluice with water. The gold collects in front of the riffles. Mercury is sometimes used to aid in collecting the gold.

Placer gold that is granular and not too fine can be recovered effectively by running the closely-sized sand over a standard concentrating table.

Several wet machines have been placed on the market recently which recover the gold from placers with a reduced amount of water. Dry placer machines, in the operation of which no water is required, do not save as much gold as wet machines, and they have not been satisfactory with many gravels. If these machines are sufficiently improved they may be used successfully at certain deposits where the water problem is serious.

Hydraulic mining consists of excavating the gravel deposit by directing a stream of water under considerable pressure against it, washing the material to a sluice through which it passes and in which the gold is ca. ... and disposing of the tailings. The most important requirements for this type of mining are an ample supply of cheap water under high pressure, sufficient grade on bed-rock for the sluices, and adequate dump room for the tailings. The disposal of the tailings may entail difficulties in agricultural areas. Under favorable conditions hydraulic mining is much less expensive than hand methods or ordingary sluicing.

The gold dredge is a scow supplied with a mechanical excavator and elevator - usually a digging ladder and an endless system of buckets, screening and washing plant, and gold-saving equipment. This method is suitable to extensive river-bar and gravel-plain placers which are fairly thick and have a level bedrock. The dredge must operate while floating on the water, and hence the conditions must be such that a pond for it can be maintained. Numerous large boulders in the gravel may seriously interfere with dredging operations Dredges are expensive, but they handle material at a lower cost per cubic yard than is possible by any other method.

MINING AND MILLING OF GOLD ORES

Lode gold ores are ordinarily mined in the same manner as the lode deposits of other metals having the same size, shape and physical characteristics. A discussion of methods is beyond the scope of this report.

Gold ores consist of (a) shipping ores, or those rich enough in gold to be shipped to a smelter, and (b) milling ores, or those too low in gold content to stand transportation and smelter treatment costs, but which can be profitably milled near the deposit and the resulting bullion or other product marketed. Some ores that are below shipping grade as mined can be hand-sorted, and a product obtained that is acceptable at the smelter. At most gold deposits the amount of milling ore is many times larger than the shipping ore. Most of the ore taken from New Mexico gold mines has been milled.

Many gold mills have been constructed and then operated for only a few weeks or months because of a deficiency in the amount or value of the ore, or both. It is never advisable to construct a mill until a sufficient quantity of ore of satisfactory grade is blocked out to justify its cost, and only after experimental work has definitely indicated the best milling process.

The most important milling methods for gold ores are amalgamatión, cyanidation and concentration. In the amalgamation process the ore is usually crushed in a stamp mill and the gold collected by mercury. This process is only adaptable to free-milling ores - those in which the gold is in the native form. This class of ore is usually confined to the oxidized zone, and the deeper sulfide ores may not be amenable to this process. the amalgam, consisting of mercury and gold, is retorted, and the residue containing the gold is melted into bars, in which form it is sold. Silver in the native form is also recovered by this process.

In the cyanide process the ore is ground to the necessary fineness and the gold is dissolved in an alkaline solution of potassium or sodium cyanide. From the solution the gold is precipitated by zinc shavings or zinc dust. The precipitate is run through a filter press and melted into bullion bars for the market. Cyanidation is effective for the recovery of both gold and silver in the native form and also when contained in the sulfides and other minerals. Copper in the ore may interfere with cyaniding.

In some mills gold ores are concentrated by gravity methods or flotation, and the concentrates are sold to a smelter. Payment is made by the smelter not only for the gold and silver in the concentrates but for certain other metals if present in sufficient amounts. For many ores some combination of the above methods gives best results.

HISTORY

It is possible that the Indians obtained small amounts of gold from the gravels of the State before the arrival of the Spaniards, and there is some evidence that the early Spanish settlers did some mining for gold and silver. However, the first recorded discovery of gold was made in 1828 when the Old Placers in the Ortiz Mountains south of Santa Fe were opened. This discovery inaugurated the era of placer mining in New Mexico.

Though these rich placers were worked by the most primitive methods, a great deal of gold was taken from them in the next ten

years. In 1839 the still richer New Placers at the foot of the San
Pedro Mountains were discovered. The production soon decreased, but
it is safe to say that more or less placer gold has been obtained
each year since 1828 from one or both of these districts. Placer
gold was found in the Elizabethtown district, Colfax County, Pinos
Altos district, Grant County, and the White Oaks district, Lincoln
County, in the sixties, and these and other placer districts yielded
considerable gold prior to the important lode mining of the eighties.

Lode mining for gold began in 1833 when the Ortiz gold quartz
vein was discovered near the Old Placers. Aside from operations in
this district, lode deposits received but little attention until
1860-1870, when discoveries were made in Colfax, Grant, Lincoln,
and other counties. The Civil War and the later depredations of
the Indians retarded development, and important gold lode mining
operations did not begin until 1885. The period of most active lode
mining for gold was from that year to the beginning of the present
century.

PRODUCTION

The production of gold from lodes and gravels in New Mexico
prior to 1880 was about 749,000 ounces, valued at $15,483,000; from
1880 until 1904 the output is estimated at 654,717 ounces, valued at
$13,534,219; and official records show that from Juanuary 1, 1904
until December 31, 1939, the amount produced was 1,117,273 ounces,
valued at $26,304,707; making an estimated total for the State of
2,520,990 ounces, valued at $55,321,926. Placer mi have yielded
$12,000,000 to $15,000,000.

Similarly, in the 9 year period from January 1, 1930 until
December 31, 1938, lode and placer mines produced 291,213 ounces
of gold, the value of which was $8,698,382. Of this amount 102,030
ounces or 35 percent, was yielded by dry and siliceous gold, gold-
silver, and silver ores; 13.7 percent came from copper and copper-
lead ores; 44.9 percent from lead, lead-zinc, and zinc ores; and
6.3 percent came from placer mines.

These figures show a marked increase in placer gold production.

GENERAL FEATURES OF THE DEPOSITS

Placer Deposits

Gold placer deposits consist of sand, gravel or other detrital
rock containing gold in commercial amounts. The material is usually
called gravel, and the deposits are known as placers.

The formation of placers is dependent on the occurrence of gold
in lode deposits in rocks which are subject to erosion; the freeing
of the gold from the rock by weathering or abrasion; and the transportation, sorting and deposition of the gold-containing detrital
material resulting from erosion. Original placers may be reworked by
later stream action and new placers formed.

Workable placers may result from the erosion of lode deposits too poor for profitable mining, and high-grade lode deposits in which the gold is finely divided may not yield placer deposits. Placers may have come from high-grade primary deposits now entirely removed by erosion.

In most placers the greatest concentration of gold occurs at or near bedrock. Impervious beds within the gravels may have gold concentrated just above them and are spoken of as "false bedrock". The rich gravels usually are in ribbon-like "pay streaks" which may or may not follow the course of the present drainage channels. The gold varies in size from nuggets to minute flakes called "colors". Fine, flaky gold is difficult to save. Placer gold is usually accompanied by "black sand", consisting of magnetite, ilmenite, hematite, garnet, zircon, and other heavy minerals. Because of their high specific gravity, these minerals usually collect with the gold during concentration, even though not closely associated with it in the gravel. Some "black sands" contain gold which cannot be separated from the minerals mechanically, in which case the sands may be marketable at a smelter. New Mexico placer gold is mostly above .900 fine.

Placers may be classified as (a) residual or alluvial placers when formed directly over the outcrops of the lode deposits; (b) hillside placers on valley slopes, which are partly sorted by running water but not in distinct channels; (c) gulch or creek placers which are shallow placers in or adjacent to the beds of small streams; (d) bench placers or terrace gravels, consist' of old stream gravels partly removed by later streams which have cut into the original bedrock; (e) river-bar placers, which occur in river bars and in gravel flats adjacent to larger streams of small gradient; (f) gravel-plain placers formed in flood plains, deltas and alluvial fans; and (g) buried placers, which have been buried by a later accumulation of sediments or by surface flows of igneous rock.

The New Mexico placers include most if not all of the above types. In general they are shallow, 50 feet being an unusual depth. In the Elizabethtown district some deep gravels are reported, but they have not been explored. The placers are largely of Quaternary age, but in places, as in the vicinity of the San Pedro Mountains where erosion has been active with little interruption since Tertiary time, some of the lower gravels may have accumulated in that period. No ancient well-defined buried channel systems of placers, like those in California, have been discovered, but they may possibly occur.

The principal New Mexico placer districts are the Old Placers at Dolores, and the New Placers at Golden, both in Santa Fe County; the Elizabethtown or Moreno placers and smaller placers on Cimarroncito and Ponil Creeks, Colfax County; the Hopewell placers, Rio Arriba County; the Hillsboro (Las Animas) placers, and the Pittsburg placers near Shandon, Sierra County; the Nogal and Jicarilla placers, Lincoln County; and the Pinos Altos placers, Grant County.

Lode Deposits

Gold or silver, or both, are contained in the ores of nearly every mining district in New Mexico. In many of them the precious metals are subordinate to copper, lead, or zinc, but there are a number of districts whose ores are valuable chiefly for their contained gold. Only those districts in which gold is especially important are considered in this report.

Deposits connected with Tertiary Intrusive Rocks. - Many of the gold lode deposits of the State are connected with stocks, dikes, and other intrusions of monzonite and related rocks of Tertiary age. They occur chiefly in veins, shear zones and brecciated zones in the intrusive rocks, but in part as veins and replacement deposits in the adjacent sedimentary or igneous rocks. In places the invaded rocks have been greatly metamorphosed by the intrusions, and the deposits in them are essentially contact-metamorphic deposits. The gold occurs chiefly in the native form, in pyrite, and other sulfides, and sparingly as the telluride. Silver is present in varying amounts as argentite, born silver, alloyed with native gold, and as an impurity in minerals of the other metals. One or more of the sulfides, pyrite, chalcopyrite, bornite, galena, and sphalerite, are usually present in small or moderate amounts, and in some deposits the base-metal sulfides are of economic importance. In places near the surface the sulfides have been changed to limonite, copper carbonates, smithsonite, cerussite, and other oxidized-zone minerals. In some deposits oxidation has extended to considerable depths. Quartz is the chief gangue minerals in a places it is accompanied by calcite, fluorite, and barite. Placers have resulted from the erosion of these deposits.

Deposits in Tertiary Extrusive Rocks. - Surface flows of rhyolite, latite, andesite, and related rocks contain a number of important gold deposits. These flows occupy large areas and are thousands of feet thick in places. The deposits have no apparent connection with intrusive igneous rocks. They consist of veins, many of which are short and irregular. The mineralogy of the deposits is quite similar to that of the deposits connected with Tertiary intrusive rocks, but ruby silver and related silver minerals occur. Much of the gold is so finely divided that no colors can be obtained by panning. In general, copper, lead, and zinc are present in very small amounts, and the veins consist chiefly of gangue minerals. As a rule these deposits do not yield placers.

Deposits in Precambrian Rocks. - Gold lode deposits of minor importance occur in the Precambrian rocks. Some of these deposits may have formed in Precambrian time, but others are doubtless more recent. They consist of veins, shear zones, and disseminated deposits in schist. The minerals are much the same as in the deposits connected with Tertiary intrusive rocks, but copper is relatively more abundant, zinc is present in moderate or small amounts, and lead is usually absent. The gangue of veings is mainly quartz, which is accompanied by calcite, siderite, tourmaline, and specularite. The gold of these deposits frequently collects in placers.

AREAS FAVORABLE FOR NEW DISCOVERIES

The areas considered most promising for prospecting for placer deposits are those containing gravels which have been eroded from the known gold lode districts, and particularly those which have yielded placer gold in the past. These areas are specified or indicated in the discussion of the districts in the latter part of this report. If the older rocks contain no gold and give no evidence of primary metallization, the gravels derived from them are almost certain to be barren.

Additional gold lode deposits of value no doubt will be found from time to time. In general, the areas near the proved districts which have similar geologic features offer the most promise for new finds. Other areas in which Tertiary intrusive rocks are prominent deserve attention, and the Precambrian areas are moderately promising. Additional discoveries are possible in the thick Tertiary extrusive rocks, but large areas of these rocks are undoubtedly barren. The most favorable parts are those consisting of thick-bedded brittle flows.

The known mineralized portion of the State is a broad zone which extends from the north to the south boundary and includes Taos, eastern Rio Arriba, western San Miguel, Santa Fe, Sandoval, eastern Bernalillo, eastern Valencia, western Torrance, western Lincoln, Socorro, southern Catron, Sierra, Otero, Dona Ana, Luna, Grant, and Hidalgo Counties. The north-central part of Valencia County, although not in this ·e, is also mineralized. Future gold discoveries are more probable in these areas than in the remainder of the State.

CATRON COUNTY

Mogollon (Cooney) District

This district is in the southwestern part of Catron County and about 85 miles by good road from Silver City. It was discovered in 1875, but active production did not begin until the nineties. The deposits occur as veins in Tertiary rhyolite, andesite, and latite, and contain gold, silver, and copper. The total production of the district has been about $20,000,000, approximately one-third of which represents gold. The principal mines were closed down in 1925, but Ira L. Wright and associates of Silver City have recently leased and resumed mining and development work at the Fanny mine.

Gravels of the Mogollon district have yielded little, if any, placer gold.

COLFAX COUNTY

Elizabethtown District and Vicinity

Placer gold was found near what is now Elizabethtown, Colfax County, in 1866, and in the following spring there was a great influx of miners to the district. Most of the early work was in gravels, although some lode deposits, including the famous Aztec mine, were located. The total production of this region has been about $6,000,000, over 99 percent of which was derived from lode and placer gold. Lode deposits have yielded a little over 55 percent of the total gold. About 1900 production rose to $100,000 a year in consequence of dredge operations, which were soon discontinued.

The lode deposits occur as quartz veins and as contact-metamorphic bodies at the contact of the Raton conglomerate with the underlying Pierre shale. Most of the ore has been taken from the shale in which it generally occurs at the crest of minor folds.

The placer deposits occur mainly on Moreno, Ute, and Ponil Creeks and most of the production has come from the Moreno drainage area. The gold undoubtedly has been supplied largely by the eroded part of the Aztec lode. In 1869 water was first received at these workings through the "Big Ditch". This ditch was 41 miles long and brought water from the headwaters of the Red River, 11 miles west of Elizabethtown.

The placers in this region vary in thickness from a few feet to over 300 feet, but for the most part they are confined to narrow valleys where the steep gradient has not permitted the building up of thick deposits. These placers have been worked by dredges, hydraulicking, ground sluicing, and simple hand methods. In places the gravel has been removed to bedrock, but in others considerable yardage remains.

The "Big Ditch" is no longer kept in repair, and most of the water now used comes through relatively short ditches from sources closer to the gravels being worked. Reservoirs are sometimes used to accumulate enough water for a few days work. Recent operations have been sporadic and comparatively unimportant, but in periods of unemployment miners have reworked the gravels, earning a few dollars a day.

The gravels are on the Maxwell land grant, and information concerning their availability for lease or purchase can no doubt be obtained from the Maxwell Land Grant Co., Raton, New Mexico.

DONA ANA COUNTY

The mines of Dona Ana County, from the first recorded production to the end of 1938, produced #127,342 worth of gold, most of which was derived as a by-product from the base-metal ores of the Organ district.

Black Mountain and Texas Creek Districts

In the Black Mountain and Texas Creek Districts small amounts of gold occur associated with pyrite and other sulfides in quartz veins. The gold is free only near the surface. A small production has been reported.

GRANT COUNTY

Gold Hill District

This district lies 16 miles by road northeast of Lordsburg in Grant County. Veins of massive quartz occurring in Precambrian granite and associated with pyrite or its oxidation product, limonite, were discovered here in 1884. Two stamp mills were erected and operated on $15 to $40 ore until the apparent exhaustion of the oxidized ores. There is very little activity in this district at present.

Malone District

The ores of the Malone district were discovered in 1884, but placer mining had been carried on in this vicinity for several years prior to the lode discovery. The district is in the southern part of the Burro Mountains, Grant County, and is a few miles north of Gold Hill.

The lode deposits occur as quartz veins along a fault contact of lava and granite. The veins extend into the granite but show little tnedency to cross the fault into the lava. Oxidized gold ores have furnished the principal production to date, but sulfides of iron, lead, and zinc occur on the deeper levels.

Pinos Altos District

This district is located in the Pinos Altos Mountains, Grant County, about 8 miles northeast of Silver City. Placer gold was found here in 1860, and later in the same year the Pacific vein was discovered. The district has yielded over $8,000,000 in gold, silver, copper, lead, and zinc. Probably 30 to 40 percent of this amount came from gold along, of which 25 to 30 percent was recovered from the gravels. In the early years of activity placers probably furnished more than one-half the gold. Several small-scale operations have been carried on during recent years.

The lode deposits are fissure veins in intrusive rocks and re-placement deposits in limestone. The veins contain chiefly pyrite, chalcopyrite, gold and silver in a quartz gangue.

Working of the gravels is hampered by the intermittent water supply, and placer production has been derived to a large extent from small operations by individuals. Bear Creek Gulch and Rich Gulch, a tributary to it, on the north, Whisky Gulch on the east, and the gulch heading near the old Gillette shaft were the principal producers. Flood waters from heavy summer rains occasionally work over the gravels, causing a reconcentration of the gold into workable deposits.

According to tests made by the United States Geological Survey in 1905, the gravels contain as much as 40 percent "black sands". One sample of "black sand" containd 83 percent magnetite, 3 percent garnet, 8 percent hematite, and $9.30 a ton in gold. Since 800 pounds

of this sand was recovered from a ton of gravel, each ton of the
original gravel contained about $3.70 in gold locked up with the
heavy minerals. Another sample of "black sand", 200 pounds of
which were recovered from a ton of gravel, assayed $13.23 a ton,
indicating a value of $1.32 for each ton of original gravel. Still
another sample showed less than one cent to the ton. All values
have been calculated at the old price of $20.67 per fine ounce of
gold.

Steeple Rock District

This district is in western Grant County near the Arizona line
and is about 20 miles northeast of Duncan, Arizona, by road. Ore
was first discovered about 1880 and the camp had an active life of
about 15 Or 20 years. Recent operations have been important,
notably on the East Camp and Carlisle groups. The Carlisle was
long the most important mine of the district and is said to have
produced $3,000,000, mostly from gold.

The ore deposits consist of quartz veins in extrusive Tertiary
rocks. Pyrite, sphalerite, chalcopyrite, galena, and some calcite
occur with the quartz. The best ore has been found within 300 feet
of the surface.

HIDALGO COUNTY

Kimball (Steins Pass) District

This district lies close to the Arizona line north of Steins
Pass station on the Southern Pacific Railroad. It was discovered
in 1875, but active mining did not begin until about ten years
later. Silver was the chief product, but considerable gold was
obtained.

The ore deposits of the Federal group apparently are richer
in fold than those of any other property in the camp. The deposits
are not true veins but are zones of brecciation in the country rock,
which is silicified rhyolite or diorite prophyry. The breccia zones
are greatly silicified and outcrop prominently. They vary in width
from 5 to 20 feet.

Sylvanite District

The Sylvanite district is about 12 miles southwest of Hachita
in the central part of the Little Hatchet Mountains. Copper was
discovered in this region in the early eighties, but it was not
until 1908 that gold was found.

The lode deposits occur as quartz veins in monzonite, in
syenite dikes, and in sedimentary beds. The vein filling within
about 60 feet of the surface consists of quartz, calcite, oxidized
iron and copper minerals, free gold, and some tetradymite. Below
this depth low-grade gold-bearing sulfides appear. In some places
copper is present in sufficient quantities to be of value.

Placer gold has been recovered from most of the gulches on the west side of the Little Hatchet Mountains and was discovered before the search for gold lodes began. Placer mining did not continue very long, and the total production from gravels is estimated at about $2,500.

LINCOLN COUNTY

Jicarilla District

The Jicarilla district is located in the central part of Lincoln County about 7 miles southeast of Ancho. It is in a group of hills known as the Jicarilla Mountains, at the north end of the Sierra Blanca. It was said that placer mining was carried on as early as 1850, but no lodes were found until the eighties. Production from placers amounts to about $145,000, and some low-grade gold ore has been treated.

The lodes occur as pyrite-quartz veinlets in small fractures in the intrusive monzonite porphyry. The country rock is somewhat silicified and impregnated with pyrite, and in some places it carries gold over a width of 40 feet or more. The gold occurs in the pyrite, and copper and silver are found with the gold in some mines.

The palcers were derived from the lodes, and the gold has not been transported very far. In places they gradually merge down ...d into undecomposed prophyry carrying some gold. Along the bedrock are streaks of good pay dirt, but they are usually overlain by several feet of overburden containing little if any gold. Water is scarce. About 40 years ago a large dredge was erected on Ancho Creek, but on account of depth to pay gravel and lack of water it was unsuccessful and was operated for only a short period. Since that time most of the placer mining has consisted of the intermittent small-scale activities of individuals. Some machinery has been installed since the price of gold went up and placer production has increased.

Nogal District

The Nogal mining region in Lincoln County comprises several small districts scattered over the Sierra Blanca southeast of Carrizozo. Vera Cruz, Parsons (Bonita), Schelerville (Church Mountain), and Alto (Cedar Creek) are some of the names that have been applied to the camps in this region. Placer gold was found in Dry Gulch northeast of Nogal Peak about 1865. A lode was located in 1868, but active prospecting did not begin until after the area was taken from the Mescalero Indican Reservation in 1882. The output of the district has been estimated at about a quarter of a million dollars.

There are two types of lode deposits in the district. One, which includes the American, Helen Rae, and other mines, consists of gold-bearing sulfides of copper, lead, and zinc in a gangue of

quartz, calcite and dolomite. The vein on which these mines are
located occurs for the most part in the monzonite porphyry and is
3 to 5 feet thick. The other type occurs at the Hopeful and Vera
Cruz mines. This deposit is simply altered porphyry, much bleached
and kaolinized. It is said to be 900 feet long, 120 feet wide,
and to reach a known depth of 260 feet. The gold content of this
material is small.

White Oaks District

This district, which is in Lincoln County, centers about
Baxter Mountain approximately 10 miles northeast of Carrizozo.
In 1879 lode gold was discovered at what became known as the
North Homestake mine and soon thereafter the Old Abe, South Homestake, and other claims were located. A small intermittent production of placer gold had been made for 25 years prior to the
lode discovery. The total production of the camps is estimated
at about $3,000,000.

The gold deposits are narrow stringers or wider lodes in a
fine-grained monzonite which is intruded into shale. Some of the
shale has been mineralized and forms ore bodies. Tungsten minerals,
fluorite and gypsum are associated with the quartz and auriferous
pyrite. The Old Abe mine was worked to a depth of 1,380 feet, but
the richest ore occurred near the surface in the form of high-grade
pockets and shoots. The ore that was mined had a value of about
$20 a ton.

OTERO COUNTY

Jarilla (Orogrande, Silver Hill) District

The Jarilla district is situated in an isolated group of
hills known as the Jarilla Mountains in Otero County, and is about
50 miles northeast of El Paso, Texas. Some work was done about
1880, and prior to 1904 the district had produced approximately
$100,000 in gold and copper, about $8,000 of which was recovered
from dry placers. More extensive metal-mining operations began
about 1900 after the discovery of turquoise in the district had
brought more miners to the area. The district has produced over
$375,000 in gold.

The lode deposits occur as contact-metamorphic bodies in the
limestone near its contact with the intrusive mass of monzonite.
The gold is associated with iron and copper sulfides, and the ores
contain about 3 ounces of silver to every ounce of gold.

The placer deposits lie on the southeastern slope of the
Jarilla Hills east of the Nannie Baird mine. Water is very scarce
and in the early days was shipped into the camp in tank cars.
Most of the placer gold has been recovered with some form of dry
washer. The gravels are reported to contain about $1 a cubic
yard. "Black sand" constitutes approximately one percent of the
gravel. A sample of this sand tested by the United States Geo-

logical Survey in 1905 was found to contain 10 percent magnetite, 1.5 percent ilmenite, 11 percent hematite, 1 percent zircon, 56 percent quartz, and $40 a ton in gold. If this sand is uniformly distributed through the pay gravel and is uniform in content, the figures above indicate that each ton of gravel contains 40¢ in gold associated with the "black sand". Another sample of "black sand" from this district contained $377 a ton in gold but only 2.5 lbs. of this sand was recovered from a ton of gravel. According to this sample, the original gravel contains about 45¢ a ton in gold.

RIO ARRIBA COUNTY

El Rito District

The so-called El Rito district is in Rio Arriba County and about 4 miles north of the town of El Rito. It is in the Chama Basin and largely between El Rito Creek and Arroyo Seco. The district has a length from north to south of about 10 miles and a width of about 4 miles.

The gold deposits of the district consist of conglomerate and minor sandstone beds resting on older formations. These beds, which are essentially horizontal, attain a maximum thickness of at least 1,000 feet, and are hundreds of feet thick in much of the area. The conglomerate is well cemented and in large part is colored a bright red by the iron oxide, hematite, Gold is very sparingly distributed through parts of the conglomerate, commonly in the fine or cementing material.

Several years ago the Arriba Gold Fields, Ltd. was organized, ostensibly to work these deposits. The company was capitalized for $25,000,000, and stock was issued on this basis with a par value of $1.00 per share. Considerable stock was sold at 50¢ and 75¢ a share.

Advertising matter which was circulated among possible investors claimed that the property contained up to 44 billion tons of ore. The grade of this ore was placed at from $3.00 to $8.00 a ton in gold, and it was claimed that the ore also contains 2½ ounces of silver and 3 pounds of mercury to the ton. The total gross value of the deposits in gold alone was estimated at 264 billion dollars by one especially optimistic investigator of the deposits. It was stated that "it is a virgin property that has never been defiled by the hands of man". Favorable reports were submitted by five men who were at least inferred to have the degree of "Engineer of Mines"; also by a "consulting engineer" and a graduate civil engineer.

The stock selling activities of this company were brought to the attention of Mr. L.A. Tamme, of the Blue Sky department of the office of the New Mexico Bank Examiner. At his request the property was examined briefly by Mr. C.G. Staley, state geologist, and Mr. W.C. Powell, geologist for the State Engineer. Seven samples were

taken by these men, some of them from the same exposures that
were sampled by men reporting favorably for the company, which
showed an average of 10¢ a ton in gold, a trace of silver (of
no value), and no mercury. The highest grade sample taken by
Staley and Powell contained 30¢ a ton ingold. The material
sampled is so well cemented that it would require blasting and
crushing prior to treatment. The deposit was considered by
Staley and Powell to offer no possibilities for commercial mining.

Hopewell District

The Hopewell district is a westward extension of the Bromide
district and lies in an area of Precambrian rocks about 15 miles
west of Tres Piedras, Taos County, and about 25 miles south of
the Colorado line. Placer deposits were found the same year.
The district has produced over $300,000 in gold, approximately
95 percent of which was recovered from placers.

The lode deposits consist of quartz veins and fahlbands
occurring generally in the schist. The veins are only a few inches
wide but occasionally are found close enough together to form a
lode. Pyrite is the most common sulfide, but some chalcopyrite is
found. The gold occurs in pyrite and in the oxidized zone as
free gold.

It is reported that $175,000 in gold was recovered from placers
during the first three years of mining. A few nuggets worth from
$30 to $96 were found. There are two placer areas in the district,
both on Eureka Creek. The Fairview placer lies immediately west
of the town of Hopewell. It is said that this ground was very
rich where the valley narrows to a steep-sided channel. The Lower
Flat placer is about a mile farther down the creek at the junction
of Eureka Creek with a branch of Vallecitos Creek. The gravel of
this deposit was over 35 feet deep in some places. An attempt at
hydraulic mining was made here many years ago, but the water
supply apparently was too meager for such operations. There has
been but little activity in this district in the last 35 years, the
production of placer gold for that period having been about $2,000.

SANDOVAL COUNTY

Cochiti (Bland) District

The Cochiti district in Sandoval County is in the southern
foot-hills of the Valle Mountains about 30 miles west-northwest of
Santa Fe. Some prospectors visited the area as early as 1880, but
it was not until 1894 that production began. Since that date the
district has produced approximately $1,230,000 in gold and silver,
about 70 percent of which came from gold. There has been no placer
production from this district.

The region is one of the extensive rhyolite flows surrounding
a mass of monzonite. The ores are confined to the monzonite, and
at no place are they known to extend into the rhyolite. The ore

bodies consist of quartz veins and lodes in brecciated zones.
Argentitite, presumably auriferous, is the principal ore mineral,
although some galena and chalcopyrite and abundant pyrite and
sphalerite are found. Small rich pockets of ore were common.

Placitas District

The Placitas district is in the northern end of the Sandia
Mountains about 8 miles east of Bernalillo. Some prospecting
has been done on lode deposits of lead, copper, and silver, and
a deposit of gold-bearing gravel conglomerate has been worked as
a placer mine. The only recorded production of placer gold from
Sandoval County was made in 1904 when $1,013 was reported. This
may have come from the Placitas district.

SANTA FE COUNTY

Two of the richest placer districts in New Mexico, the Old
Placers and the New Placers, were discovered by Spanish Americans
in 1828 and 1839. For many years they yielded a heavy production.
It has been estimated that the Old Placers yielded $60,000 to
$80,000 annually from 1832 to 1835, but in later years not more
than $30,000 to $40,000. These placers have yielded close to
$4,000,000 or approximately 25 percent of the placer yield of New
México and 6 to 8 percent of the total gold yield of the State.

Old Placers (Ortiz, Dolores) District

The Old Placers district is on the Ortiz Grant in Santa Fe
County. It is in the Ortiz Mountains between the San Pedro
Mountains and the Cerrillos Hills. The lode deposits consist of
quartz veins and contact-metamorphic bodies. The quartz veins
occur in brecciated porphyry. Rich ore shoots were found in the
oxidized zone, generally within 200 feet of the surface. The
contact-metamorphic ores consist of garnetized limestone through
which are scattered grains of auriferous (gold-bearing) chalcopyrite.

The placers are situated at the mouth of Cunningham Canyon
at the old town of Dolores and on Dolores Gulch and Arroyo Viejo.
The gravels form a mesa which is the upper part of an old alluvial
fan. They occupy a large area, and considerable valuable ground
is said to remain. In some places the gravels are as much as 100
feet thick. The scarcity of water is a serious handicap to profitable operations, but the gravels are too wet for strictly dry
placer methods. Many years ago Thomas A. Edison tried to work the
gravels by an electrostatic concentration process, but the dampness
of the gravels prevented economic recovery. From 2 to 50 pounds
of "black sand" are recovered from each ton of gravel, and this
sand contains from $4 to $30 a ton in gold. Recently the district
has yielded a small amount of gold each year obtained mainly by
Spanish American miners operating on a very small scale. The gold
is about .918 fine.

New Placers (San Pedro) District

This district, located in the San Pedro Mountains and partly in the Ortiz, Grant, Santa Fe County, has been one of the richest placer areas in the State and contains lode-deposits of gold, copper and lead. The ore deposits are similar in many ways to those of the Old Placers district. The veins are small but very abundant, both in porphyry and in the sedimentary rocks. Auriferous copper sulfides in a garnet gangue are found in the contact zone. The gold is distributed erratically through the veins, and although high-grade pockets have been found, no lode mine of permanent value has been developed. The depth of oxidization rarely exceeds 100 feet.

Gold-bearing gravels have accumulated at the foot of the San Pedro Mountains, especially to the north, west, and south. As the erosion of these mountains has proceeded the detritus of porphyry and limestone has extended for miles in every direction, especially westward along Tuerto Creek. All this subangular gravel is said to contain gold, and every creek and gulch cutting it has reconcentrated the gold in the stream bed.

Several unsuccessful attempts to work the placers of this district with dipper dredges have been made. The overburden varies from 10 to 40 feet in thickness. The gravels contain 10 cents or more in gold to the cubic yard, and it is though that a fairly large yardage containing $1 or more is available at bedrock. The greatest difficulty is the serious scarcity of water, for the mountains contain no perennial streams. Water is found in wells at a depth of 500 to 800 feet, and this water rises to within 100 or 200 feet of the surface. Several wells have been drilled and so far have yielded as much as 25 gallons a minute from each hole.

Along Tuerto Creek a few miles below Golden are the gold-bearing "cement-beds" which have been worked unsuccessfully a number of times. These beds rest on sandstone of the "Red Beds" series and are 50 to 100 feet in thickness. They are roughly stratified sub-angular gravel deposits consisting largely of porphyry and limestone fragments, presumably derived from the San Pedro Mountains. At least one attempt has been made at recovery by stamp-milling.

SIERRA COUNTY

Hillsboro (Las Animas) District

Hillsboro, the county seat of Sierra County, is about 17 miles north of Lake Valley and 16 miles west of the Rio Grande. Float from the vein that was subsequently located as the Rattle-snake mine was discovered in 1877, and later in the same year placer gold was discovered in Snake and Wicks Gulches. That winter one miner was said to have taken $90,000 in gold from Wicks Gulch. The total production of the district is estimated at about $7,500,000. About $6,500,000 of this production represents gold, of which nearly one-half came from placers.

The gold lode deposits occur as quartz veins in shear zones in andesite. They consist chiefly of copper sulfides carrying gold and silver and some free gold. Most of the veins are narrow but high grade.

Intermittent and small-scale lode mining has been conducted in the district for many years. Recently a number of old mines have been reopened and development and mining resumed. Most of the old workings have been under water for many years.

Dry placer mining was carried on in what is known as the Las Animas or Gold Dust district about 6 miles northeast of Hillsboro, in 1912. According to V.C. Heikes ("Dry Placers in New Mexico"; U.S. Geological Survey Mineral Resources of the United States, calendar year 1912, Part I, p. 262, 1913), the placers which were worked cover several square miles from one-half inch to several feet deep, and the gold is found on false bedrock. Heikes says that fake tests averaged $1.25 a cubic yard but working tests showed only 22¢ to 25¢ a yard.

The total amount of gold in these gravels is undoubtedly large, but most of the ground is low grade. Here, again, the water situation is serious. Several wells have been drilled and at least one was dug. Little information is at hand as to the quantity of water thus made available, except for the report that one or two of these wells will yield about 25 gallons per minute.

Pittsburg (Shandon) District

The Pittsburg district lies on the east side of the Rio Grande and on the southwest slopes of the Caballos Mountains. The known gold-bearing gravels are in sections 16, 17, 20, and 21, T. 16 S., R. 4 W., mainly along Trujillo Gulch. The gold occurs in a coarse granitic sand which is several feet deep in places. This sand was derived from the granite of the Caballos Mountains which contains gold-quartz veins. The district has produced about $175,000 in placer gold. Very little water is obtainable at the diggings.

Encarnacion Silva began working these deposits about 1900, but the discovery did not become generally known until 1903. Several companies have worked these gravels with water pumped from the Rio Grande, two miles away.

SOCORRO COUNTY

Rhyolite District

This district is in the southern end of the San Mateo Mountains about 30 miles southwest of San Marcial and about 20 miles south of Rosedale. The lode occurs along a wide fault zone in rhyolite porphyry. The values are in gold. Very little development has been done, and the workings consist of a few small prospects.

Rosedale District

The Rosedale district is in the northern part of the San Mateo Mountains about 25 miles southwest of Magdalena. The first discovery was made in1882, and after operations began in the nineties production was maintained until 1916. No production figures are available, but it is thought that gold worth about $500,000 was recovered. There was no placer production.

The ore occurs as a manganese-stained quartz in well-defined shear zones in rhyolite porphyry. The silicified outcrops stand out clearly. The Rosedale mine has reached a depth of 732 feet with levels at 100-foot intervals. Water was encountered at 726 feet. The ore is entirely oxidized, and sulfides are absent.

San Jose (Nogal) District

The San Jose district has received considerable publicity during the past year as a result of the highly interesting discovery of high-grade gold-silver ore in the outcrop of the Pankey vein. The district is in the southern part of the San Mateo Mountains and about 3 miles north of the Rhyolite district. The Pankey vein shows prominently at the surface and can be followed on both sides of Springtime Canyon for several thousand feet. The vein consists of manganese-stained quartz in rhyolite and rhyolite porphyry. The ore is chiefly valuable for its gold content, but silver is an important constituent. Other veins occur in the district, and many claims have been located.

The Nogal Mines, Inc., has initiated an important development and mining program at the Pankey vein. An excellent camp has been built, and the road to the main highway between Socorro and Hot Springs has been improved. In March, 1932, the first shipment of several cars of low-grade ore mined from the outcrops of the vein was made to the El Paso Smelting Works of the American Smelting & Refining Company. The Springtime Mining Co., owns and intermittently operates a 40-ton flotation mill in the district.

TAOS COUNTY

Since 1904 Taos County has produced $17,000 in lode and placer gold, and the total gold production probably does not exceed $75,000.

Red River District

The Red River district lies near the head waters of Red River, close to the eastern border of Taos County and about 20 miles south of the Colorado line. Prospecting for placer deposits began about 1869 and some gold was produced. At least one attempt at hydraulic mining has been made. Production has been relatively unimportant,

but small-scale operations have been carried on by native miners
nearly every year.

The gold-bearing veins consist generally of the sulfides of
silver, copper, and lead in a quartz ganque. Some tellurides have
been found. The veins are narrow and discontinuous; they occur
most commonly in the porphyry but cut the volcanic rocks in places.

Rio Grande Valley

The sand and gravel deposits of the Rio Grande valley in Taos
County have received the attnetion of a number of prospectors and
engineers. These deposits occupy an area of several hundred square
miles and are hundreds of feet thick in parts of the valley. Gold
occurs in places in these sedimentary beds. The most promising
part probably begins where Red River enters the Rio Grande in the
northern part of the county and extends along the river to the
south boundary and into Rio Arriba County as far as Embudo. The
Rio Grande flows in a moderately deep gorge for a considerable
distance in the gravel area.

The sedimentary deposits of the valley are erosional products
from a number of mountainous areas within the drainage basin. In
large part they have been derived from the Sangre de Cristo Range
of New Mexico and Colorado. This range has a number of gold lode
deposits, but in the New Mexico portion they are relatively unimportant and probably have not supplied any great amount of placer
gold. In much of the upper Rio Grande drainage basin the older
rocks give no evidence of primary gold mineralization.

Placers in this area consist of bench placers, river-bar
placers and deep river placers. The bench deposits are the most
extensive and they undoubtedly contain some gold. These sands
and gravels are capped in large areas by a recent flow of basalt,
and basalt flows of unknown extent are interbedded with the sediments. The river-bar deposits are much less in quantity than the
bench deposits, but they have yielded practically all of the small
amount of gold obtained. In places these gravels may be fairly
rich. The gravels below the level of the river are of unknown
thickness in most of the river channel, but bedrock is probably
too deep to be reached in dredging operations in many places.
Several tests of these sands and gravels are said to have obtained
gold in encouraging amounts, with the greatest value just above
bedrock.

The potential value of the sands and gravels of this part of
the Rio Grande valley can be determined only by careful and thorough
sampling, which in view of their extent would be an expensive
undertaking. Little reliance should be placed on the returns from
a few samples that may represent but a small fraction of the material
that would have to be handled in working the deposits. Hydraulicking
of the bench deposits would be feasible only if a considerable
thickness is of workable grade. Certain "pay-streaks" might be won
by drift mining,but the gold content would have to be much greater

than for surface operations. The sampling of the river-bar gravels presents no formidable problems. The deep river gravels are worthy of special consideration because of the small cost of dredging operations under favorable conditions, but the accurate sampling of these gravels would require considerable painstaking and costly work.

The sedimentary deposits of the Rio Grande valley in Taos and eastern Rio Arriba counties illustrate a number of difficulties that successful placer mining may have to overcome. Disposal of the tailings in large-scale operations would probably interfere with farming along the Rio Grande to the south and would, no doubt, meet with objections from sportsmen because of its detrimental effect on fishing in the river. The numerous large basalt boulders would undoubtedly cause complications in dredging operations along the channel, and the occasional flood stages of the Rio Grande would have to be considered.

Detailed and accurate data on all features of these and other similar deposits should be obtained and properly evaluated prior to any large expenditure of money for promotion, equipment, or operation.

BIBLIOGRAPHY

Papers on New Mexico Gold Deposits

Barbour, Percy E., The Cochiti mining district, New Mexico: Engineering & Mining Journal, vol. 86, pages 173-175, 1908.

Brinsmade, Robert Bruce, Development of San Pedro Mountain, New Mexico: Mining World, vol. 28, pages 1021-1024, 1908.

Bush, Faris F., Red River mining district, Taos County, New Mexico: Mining World, vol. 42, pages 541-543, 1915.

_____, The Steeple Rock mining district, New Mexico: Mining World, vol. 42, pages 845-846, 1915.

Carruth, J.A., New Mexico gold gravels: Mines and Minerals, vol. 31, pages 117-119, 1910.

Chase, Charles A., and Muir, Douglas, The Aztec Mine, Baldy, New Mexico: American Institute of Mining Engineers, Preprint No. 1193, 12 pages, 1922; Transactions, vol. 68, pages 270-281, map, 1923.

Day, David T., and Richards, R.H., Black Sands of the Pacific Slope (and other areas, including a few samples from New Mexico) U. S. Geological Survey, Mineral Resources of the U.S., calendar year 1905.

Dinsmore, Charles A., The New Gold Camp of Sylvanite, New Mexico: Mining World, vol. 29, pages 670-671, 1908.

Dunham, Kingsley Charles, The Geology of the Organ Mountains, with an account of the Geology and Mineral Resources of Dona Ana County, New Mexico: New Mexico School of Mines, State Bureau of Mines and Mineral Resources, Bulletin 11, 272 pages, maps, 1935.

Ferguson, H.G., Geology and Ore Deposits of the Mogollon mining district, New Mexico: U.S. Geological Survey, Bulletin 787, 100 pages, map, 1927.

Harley, G. Townsend, The Geology and Ore Deposits of Sierra County, New Mexico: New Mexico School of Mines, State Bureau of Mines and Mineral Resources, Bulletin 10, 220 pages, maps, 1934.

Heikes, V.C., Dry Placers in New Mexico: U.S. Geological Survey, Mineral Resources of the U.S., Calendar year 1912, Part I, pages 261-262, 1913.

Henderson, Chas. W., Gold, silver, copper, lead, and zinc in New Mexico: Published annually in Mineral Resources U.S. and Minerals Yearbook by the U.S. Geological Survey and the U.S. Bureau of Mines, 1908 to date.

Jones, Fayette A., New Mexico Mines and Minerals: 349 pages, Santa Fe, New Mexico, 1904.

_____, Placers of Santa Fe County, New Mexico: Mining World, vol. 25, page 425, 1906.

_____, Sylvanite, New Mexico, The New Gold Camp: Engineering & Mining Journal, vol. 86, pages 1101-1103, 1908.

_____, The New Camp of Sylvanite, New Mexico: Mining Science, vol. 58, pages 489-490, 1908.

_____, The Mineral Resources of New Mexico: New Mexico School of Mines, Mineral Resources Survey of New Mexico, Bulletin 1, 77 pages, 1915.

Keyes, Charles R., Geology of the Apache Canyon Placers (Shandon or Trujillo Placers): Engineering & Mining Journal, vol. 76, pages 966-967, 1903.

Lasky, S.G., The Ore Deposits of Socorro County, New Mexico: New Mexico School of Mines, State Bureau of Mines and Mineral Resources, Bulletin 8, 139 pages, maps, 1932.

_____, Geology and Ore Deposits of the Bayard Area, Central Mining District, New Mexico: U.S. Geological Survey, Bulletin 870, 144 pages, maps, 1936.

_____, Geology and Ore Deposits of the Lordsburg Mining District, New Mexico: U.S. Geological Survey, Bulletin 885, 52 pages, maps, 1938.

_____, and Wootton, T.P., The Metal Resources of New Mexico and their Economic Features: New Mexico School of Mines, State Bureau of Mines and Mineral Resources, Bulletin 7, 178 pages, maps, 1933.

Leatherbee, Brigham, Mesa del Oro Placer Grounds: Mining World, vol. 35, page 536, 1911.

Lee, Willis T., The Aztec Gold Mine, Baldy, New Mexico: U.S. Geological Survey, Bulletin 620, pages 325-330, 1916.

Lindgren, Waldemar, Graton, L.C., and Gordon, C.H., The Ore Deposits of New Mexico: U.S. Geological Survey, Professional Paper 68, 361 pages, maps, 1910.

Metzger, O.H., Gold Mining in New Mexico: U.S. Bureau of Mines, Information Circular 6987, 71 pages, 1938.

Paige, Sidney, The Ore Deposits near Pinos Altos, New Mexico: U.S. Geological Survey, Bulletin 470, pages 109-125, maps, 1911.

Smith, E. Percy, and Dominian, L., Notes on a Trip to White Oaks, New Mexico: Engineering and Mining Journal, vol. 77, pages 799-800, 1904.

Statz, B.A., The New Placer Mining District, New Mexico: Mining Science, vol. 66, page 157, 1912.

_____, Geology of the Cochiti Mining District, New Mexico: Mining Science, vol. 66, pages 276-277, 1912.

Stone, George H., Dry Gold Placers of the Arid Regions (New Mexico): Mines and Minerals, vol. 19, pages 397-399, 1899.

Wynkoop, W.C., The Cochiti District, New Mexico: Engineering & Mining Journal, vol. 70, pages 215-216, 1900.

Papers of General Interest

Baxter, Charles H., and Parks, Roland D., Mine Examination and Valuation: Michigan College of Mining and Technology, Houghton, Michigan, 1939. $3.50

Bernewitz, M.W. von, Handbook for Prospectors: McGraw-Hill Book Co., New York City, 1935. $3.00

Boericke, William F., Prospecting and Operating small Gold Placers: John Wiley & Sons, New York City, 1933. $1.50

Emmons, William Harvey, Gold deposits of the World, with a section on prospecting: McGraw-Hill Book Co., New York City, 1937. $6.00

PART III

I. C. 6987

JANUARY 1938

UNITED STATES
DEPARTMENT OF THE INTERIOR
HAROLD L. ICKES, SECRETARY

BUREAU OF MINES
JOHN W. FINCH, DIRECTOR

INFORMATION CIRCULAR

GOLD MINING IN NEW MEXICO

BY

O. H. METZGER

AFTER THIS REPORT HAS SERVED YOUR PURPOSE AND IF YOU HAVE NO FURTHER NEED FOR IT, PLEASE RETURN IT TO
THE BUREAU OF MINES, USING THE OFFICIAL MAILING LABEL ON THE INSIDE OF THE BACK COVER.

I.C. 6987,
January 1938.

INFORMATION CIRCULAR

DEPARTMENT OF THE INTERIOR -- BUREAU OF MINES

GOLD MINING IN NEW MEXICO[1]

By O. H. Metzger[2]

CONTENTS

[1] The Bureau of Mines will welcome reprinting of this paper provided
the following footnote acknowledgment is used: "Reprinted from
Bureau of Mines Information Circular 6987."
[2] Associate mining engineer, Metal Mining Methods Section, Mining
Division, Bureau of Mines, Tucson, Ariz.

6082

CONTENTS – Continued

CONTENTS – Continued

CONTENTS -- Continued

CONTENTS – Continued

ILLUSTRATIONS

ILLUSTRATIONS - Continued

FOREWORD

This is one of a series of circulars dealing with mining and milling operations in different mining districts in the Western States. Data on operating costs, grades of ore treated, wage scales, and haulage rates and other information on mining properties are obtained from the operators and other local sources during the course of field inspections. They are believed to be substantially correct as to conditions at the time the properties were visited but may not be in accord with facts established by later developments.

Chas. F. Jackson,
Chief Engineer,
Mining Division.

INTRODUCTION

This is one of a series of papers being published by the Bureau of Mines on gold mining in the Western States. It describes the production of gold in the principal mining districts of New Mexico. The material was gathered on a trip through the State in the summer of 1936.

ACKNOWLEDGMENTS

The author acknowledges the aid and cooperation of mine operators in all parts of the State, who willingly contributed all information requested. Geological reports of the Geological Survey and the New Mexico Bureau of Mines were drawn upon freely for information; credit to individual papers is given in the text.

HISTORY

The first white men to set foot on the territory which is now New Mexico were members of expeditions sent out by the Spanish governors of Mexico for purposes of exploration. As early as 1541, advance parties of Coronado's main expedition penetrated as far north as what is now Sierra County. Other expeditions entered the territory in 1581, 1582 and 1583; in 1605 the Province of New Mexico and town of Santa Fe were founded.

Most of the early explorers who entered this part of the Southwest journeyed up the Rio Grande River in their search for gold, silver, and precious stones. They found the country thickly inhabited by the Pueblo and kindred tribes of Indians, who, being an agricultural people, resisted the attempts of the Spaniards to develop the mineral resources of the region. Some prospecting, however, was done by the Indians under the supervision of the Jesuit Fathers and professional Spanish miners and prospectors. There were early attempts at working silver deposits, and small amounts of gold were recovered from placers. Turquoise was mined in a few places but the output was of little or no commercial importance.

In 1680 the Pueblo Indians revolted against the oppressive rule of the whites, and the Spaniards were compelled to leave the country. They were permitted to return 20 years later with the stipulation that they would never again engage in mining. Prospecting and mining ceased almost entirely until the end of the 18th century, when the copper deposits of Santa Rita were discovered. These deposits, however, were relatively unimportant until recently, when adequate transportation and smelting facilities made possible their development on a large scale.

New Mexico became a Territory of the United States at the end of the war with Mexico in 1848. The construction of the Southern Pacific and the Santa Fe and Atchison and Topeka Railroads in 1879-83 resulted in great activity and expansion. Professional miners and prospectors combed the Western States in their search for gold and silver. Many of the soldiers

detailed to guard the railroad construction gangs and the overland stage
routes spent much of their leisure time in prospecting and discovered some
important deposits. During the last 25 years of the nineteenth century
practically all of the important gold and silver districts of New Mexico
were discovered and developed.

The early miners confined their attention almost entirely to gold
and silver, because high freight rates as well as lack of reduction
facilities made it impossible to produce base metals at a profit. In 1901
the El Paso smelter was acquired by the American Smelting and Refining Co.
and was enlarged to treat copper and lead ores, as a result of which
interest in mining was revived and many of the old gold and silver mines,
which contained appreciable quantities of base metals, were reopened.

GOLD PRODUCTION

The first known records of gold production in New Mexico date from
the time the Spaniards entered the territory in 1541. Gold mining as an
industry, however, was unimportant until about 1880. Table 1 shows the
annual production of gold, silver, copper, lead, and zinc from 1904 to
the end of 1936. Figure 1 shows the areas and districts that have accounted
for practically the entire production of gold in recent years.

Until the beginning of the twentieth century, gold was by far the
most important metal produced in the State. With the enlarging of the
smelter at El Paso, Tex., to treat lead and copper ores and the building of
a zinc smelter at Amarillo, Tex., to treat zinc ores, gold ceased to be
the most important metal. Gold production increased steadily until 1916,
but it was surpassed in value by the base metals early in the century. In
1915, the peak of gold production, the value of copper, lead, and zinc was
nearly five times that of gold and silver combined and more than ten times
that of gold alone.

In 1914, 1915, 1916, and 1917 gold production exceeded $1,000,000
annually. From 1917 to 1918 production fell from $1,064,158 to $682,791,
owing to the curtailment of production in many of the gold-mining camps
of the State when gold could not be produced at a profit with the high
operating costs brought about by the World War. A steady decline in production took place from 1918 to 1921, when the low point for the century,
$196,822, was reached. This decline was due in part to continued curtail-
ment of production in the gold-mining districts and in part to the slacken-
ing in production of base-metal ores containing gold. Another period of
industrial expansion and higher prices, beginning in 1922, brought about a
steady increase in the production of base metals and a consequent increase
in gold produced from base metal ores, which persisted until 1929. Although
gold production did not show a constant increase during that period, the
general trend was upward and the peak was reached in 1929, coinciding with
the peak in base-metal production. From 1929 to 1932 production of base
metals again declined, with a corresponding decline in gold production.

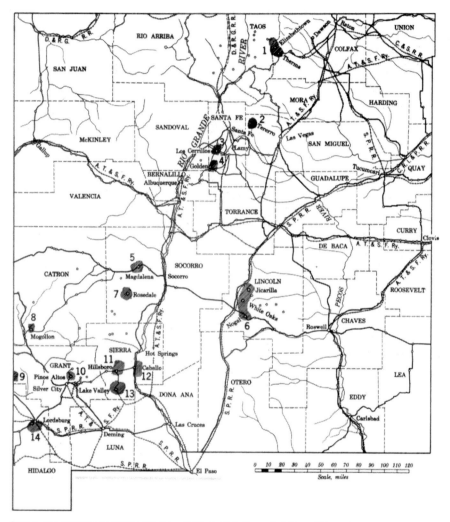

Figure 1.—Map of New Mexico showing gold-mining districts: 1, Elizabethtown, Mt. Baldy, and Therma; 2, Tererro; 3, Los Cerrillos; 4, Golden; 5, Magdalena or Kelley; 6, Jicarilla, White Oaks, and Nogal; 7, Rosedale; 8, Mogollon; 9, Steeple Rock; 10, Pinos Altos; 11, Hillsboro; 12, Pittsburg Channing; 13, Lake Valley; 14, Pyramid.

TABLE 1. - Gold, silver, copper, lead, and zinc produced in New Mexico, 1904 to 1936.[1]

Year	Gold Fine ounces	Gold Value	Silver, fine ounces	Copper pounds	Lead, pounds	Zinc, pounds	Total value
1904	18,478	$381,930	214,553	4,970,170	3,122,872	13,493,835	$1,196,390
1905	15,361	317,510	369,192	6,126,025	1,510,209	15,142,254	2,460,535
1906	14,176	293,019	491,127	7,028,670	2,987,369	17,292,655	3,203,739
1907	15,965	329,982	705,544	10,990,015	3,809,881	750,085	3,239,823
1908	14,454	298,750	405,044	6,112,630	873,763	3,567,516	1,525,091
1909	11,587	239,491	397,783	5,393,146	5,029,767	13,085,945	2,070,368
1910	23,340	482,424	842,987	4,614,386	4,320,841	18,088,129	2,691,080
1911	36,905	762,808	1,354,540	4,057,040	2,966,222	10,237,176	2,704,843
1912	37,951	784,446	1,536,701	34,030,964	5,494,018	13,566,637	8,527,955
1913	42,668	881,926	1,631,273	56,308,706	3,946,304	16,523,161	11,694,602
1914	56,687	1,171,696	1,777,445	59,307,925	1,763,641	18,403,392	11,049,932
1915	70,688	1,461,105	2,007,531	76,788,366	4,542,361	25,401,064	19,279,468
1916	66,884	1,382,480	1,765,274	92,747,289	8,214,189	36,570,649	30,827,767
1917	51,484	1,064,158	1,394,365	105,568,000	8,108,804	30,200,000	34,896,936
1918	33,033	682,791	782,421	98,264,563	8,399,239	24,050,324	28,521,413
1919	31,721	655,656	837,418	51,150,541	2,886,513	7,594,644	11,814,958
1920	23,237	480,302	768,042	54,400,691	2,869,525	10,013,580	12,367,857
1921	9,521	196,822	571,899	14,267,338	678,601	228,000	2,651,145
1922	19,964	412,693	752,240	31,537,207	3,012,223	4,496,806	5,898,446
1923	26,689	551,713	747,127	61,356,802	3,823,427	16,496,000	11,573,805
1924	24,803	512,735	795,070	74,691,436	3,634,511	20,759,200	12,470,119
1925	26,561	594,073	735,124	76,427,825	6,420,060	18,492,300	13,875,960
1926	19,630	405,803	450,934	81,642,379	6,960,366	24,104,800	14,481,808
1927	29,241	604,483	890,083	74,251,863	16,052,855	59,603,000	15,662,076
1928	32,912	680,360	827,792	89,854,646	15,610,501	62,406,000	18,815,863
1929	35,176	727,162	1,122,546	97,717,262	22,260,811	68,910,000	24,473,675
1930	32,370	669,156	1,107,335	65,150,100	20,755,900	65,529,000	13,748,217
1931	31,165	644,160	1,041,859	61,503,100	22,537,900	55,732,000	9,494,766
1932	23,208	479,753	1,142,351	28,419,000	20,227,000	51,186,000	4,734,683
1933	26,474	675,678	1,181,580	26,947,000	22,086,000	61,848,000	6,229,637
1934	27,307	954,380	1,061,775	23,630,000	18,729,000	53,043,000	6,505,002
1935	33,435	1,170,225	1,061,902	4,505,000	14,578,000	44,252,000	4,831,590
1936	33,298	1,165,430	1,141,000	6,618,000	14,307,000	41,551,000	5,346,977
Total	996,373	22,115,100	31,910,858	1,496,777,985	282,513,773	922,621,152	358,871,926

1/ Mineral Resources of the United States and Minerals Yearbook.

6082

The parallelism of the curves for gold production and base-metal production (see fig. 2) from 1917 to 1933 is due to the fact that during that period from 75 to 90 percent of the total gold was produced from complex ores containing copper, lead, zinc, gold, and silver that were mined principally for their base-metal content. In 1930, due to unemployment in other lines, many men took to prospecting and working placer deposits. Production from this source has been very small but showed a constant increase from 1930 to 1936.

In September 1933, producers of domestic gold gained the advantage of a world price of about $30 per ounce, and early in 1934 the price per ounce was fixed at $35. Interest in gold mining was revived, and many of the mines in the gold-producing areas that had been idle since 1917 and 1918 were reopened and operated. The Mogollon and Cooney districts in Catron County and the Hillsboro and Pittsburg districts in Sierra County have accounted for most of the increase in production. From 1932 to 1936, the gold-production curve showed a constant increase, whereas the base-metal curves showed a constant decrease.

Table 2 shows production of gold by counties from 1929 to 1935, inclusive, and table 3 shows production according to ore classification from 1929 to 1936. The following table shows the gold, silver, and copper content of ore from Hidalgo County and Lordsburg in terms of recovered values, 1929-32, inclusive.

Year	Gold		Silver		Copper [1]	
	Oz. per ton	Value	Oz. per ton	Value	Percent	Value
1929	0.118	$2.36	1.83	$0.98	2.3	$7.03
1930	.138	2.76	1.85	.71	2.3	6.03
1931	.121	2.42	1.44	.42	2.1	3.77
1932	.088	1.76	2.23	.63	2.6	3.28

[1] Value of copper content based on market price of copper.

In 1931 and 1932, a major portion of the ore from the Lordsburg area was classed as gold-silver ore, although the ratio of gold and silver to copper was practically the same in those years as formerly. The change in classification was due to the fall in the price of copper, which made the copper content of the ore less valuable than the gold-silver content.

Gold production of the State from gold ores, gold-silver ores, and placers shows a constant increase from 1929 to 1936, and the production from lead, zinc, and copper ores shows a nearly constant decrease. In 1935 approximately 47 percent of the total production was from ores of the first class and 53 percent was from base-metal ores.

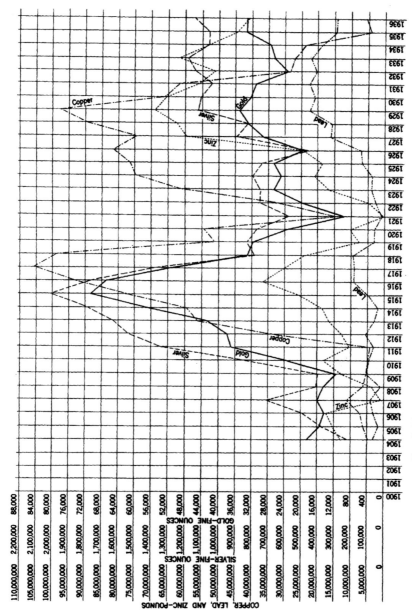

Figure 2.—Gold, silver, copper, lead, and zinc production in New Mexico from 1900 to 1936.

TABLE 2. – New Mexico gold production, in fine ounces,
by Counties from 1929 to 1935[1]

	1929	1930	1931	1932	1933	1934	1935
Catron........	738.80	931.40	733.35	3,211.05	2,651.77	3,198.57	6,946.17
Colfax........		12.00	40.05	233.08	347.14	1,146.51	2,316.88
Dona Ana.....		150.40	23.00	0.30	9.48	8.04	--
Grant.........	6,543.20	3,832.30	3,655.85	2,201.45	1,792.00	3,049.07	2,836.43
Guadalupe....	--	--	--	--	--	--	--
Hidalgo......	10,608.40	14,012.65	11,278.00	1,412.00	111.12	675.33	992.86
Lincoln......	1.21	6.23	43.40	478.15	703.47	1,041.66	893.40
Luna.........	12.00	14.80	--	1.30	.29	.92	6.20
Otero........	355.95	19.57	48.33	142.10	197.42	129.93	258.69
Rio Arriba...	--	--	--	19.90	11.95	15.11	9.91
Sandoval.....	.92	--	--	364.25	158.83	110.47	20.20
San Miguel...	16,578.20	13,102.35	14,924.55	13,941.20	19,424.69	15,632.39	14,816.66
Santa Fe.....	148.07	75.40	106.65	188.75	338.98	234.42	201.00
Sierra.......	139.33	175.80	322.50	898.95	716.53	1,894.62	3,578.63
Socorro......	14.00	5.40	--	77.65	6.92	161.00	550.89
Taos.........	--	--	9.00	14.00	8.71	8.27	7.08
Torrance.....	--	--	--	--	.29	--	--
Valencia.....	--	--	--	--	--	--	--

1/ Mineral Resources and Minerals Yearbooks, 1929 to 1935.

TABLE 3. – Gold production in New Mexico from 1929 to 1936
showing production according to ore classification.[1]

Year	From gold ores, gold-silver ores, and placers (fine ounces)	From lead zinc and copper ores (fine ounces)	Total (fine ounces)
1929	1,057.31	34,119.15	35,176.46
1930	1,941.54	30,428.88	32,370.42
1931	11,767.43	19,393.81	31,161.24
1932	6,488.08	16,722.97	23,208.05
1933	5,891.91	20,582.18	26,474.09
1934	8,600.65	18,706.36	27,307.01
1935	17,913.03	15,521.97	33,435.00
1936	19,697.06	13,339.94	33,037.00

1/ Mineral Resources and Minerals Yearbooks, 1930
to 1937.

ECONOMIC AND INDUSTRIAL FEATURES

New Mexico (fig. 1) has an average length from north to south of about 360 miles and an average width from east to west of about 340 miles, the area being 122,503 square miles. The population was 423,317 in 1930. Albuquerque, Santa Fe, and Roswell are the three largest towns, with populations in 1930 of 26,570, 11,176, and 11,173, respectively.

Two transcontinental railroads cross the State from east to west -- the Southern Pacific system in the southern part and the Atchison, Topeka, and Santa Fe system in the northern part. One branch line of the Santa Fe runs down the Rio Grande valley from Albuquerque to El Paso and another from Clovis south through Carlsbad to Pecos, Tex. Spurs from the two systems serve many of the more important coal and metal mining centers such as Silver City, Santa Rita, and Dawson.

Agriculture, stock raising, coal mining, and metal mining are the principal industries. In recent years a number of producing oil wells were brought in in the northwestern part of the State, but oil production has not yet developed into a major industry.

The mineral resources of the State are large. The coal fields of Gallup are among the most productive in the west and for many years have supplied fuel for most of the railroads and smelters of the southwest. The manganese deposits at Silver City and the iron deposits at Fierro have been, until recently, the important sources of supply of manganese and iron ores for the iron smelter at Pueblo, Colo. Although many of her lead and zinc mines have been closed since 1931, New Mexico has maintained a position of prominence in lead and zinc production throughout the years of general industrial inactivity. The Chino open-pit copper mine at Santa Rita is among the largest producers of copper in the United States. Besides this, considerable copper is produced from the complex ores from the mines of Hanover, Pecos, and Lordsburg. One of the three largest known molybdenum deposits in the world is near Questa. This deposit is of much higher grade than other deposits in the United States and has been in production for a number of years.

TOPOGRAPHY

The characteristic physical feature of New Mexico is the main branch of the Rocky Mountain range running north and south through the State. The Sangua de Cristo and the San Juan ranges of southern Colorado nearly converge in the north central part of the State, the Sangua de Cristo range forming the east and the San Juan range the west watershed of the Rio Grande River valley.

In the vicinity of Santa Fe the mountains diverge. One branch bears southwest forming the San Mateo and Cebolleta mountains just east of Gallup and terminates in the Mogollon mountains and the Black Range in the southwestern part. The other branch bears due south, forming the east watershed of the Rio Grande River and terminating in the Sacramento Mountains and the Guadalupe Mountains near the Texas border.

The Rio Grande River flows from north to south throughout the length of the State between these two ranges. At the Colorado border, the elevation is about 7,500 feet, and at El Paso, where it flows out of the State, forming the boundary between Texas and Mexico, the elevation is 3,762 feet.

The Continental Divide roughly parallels the west range. For the most part, it is to the west of the crest of the range and in many places it is in comparatively flat country. For several hundred miles in the northern part of the State it is only from 5,000 to 6,000 feet above sea level, and near Lordsburg, in the southern part of the State, it is only 4,200 feet above sea level. The only place in the State where it forms the crest of a mountain range is in the Black Mountains just east of the Mogollon Mountains.

The eastern part of the State is comparatively level, with no important mountains. The Pecos River, with headwaters just east of Santa Fe, flows south through the eastern and southeastern part of the State into Texas. The Canadian and North Canadian Rivers drain the northwestern part of the State and flow east into Texas.

The lowest point in the State is in the southeastern corner on the Pecos River, 3,000 feet above sea level. There are several peaks from 11,000 to 12,000 feet above sea level, but only one over 13,000 feet. Much of the mountainous country north of Santa Fe is from 7,000 to 8,000 feet above sea level, and small areas in the Mogollon and Black Mountains are above 7,500 feet; the rest of the State, with the exception of the mountains east of the Rio Grande River, is comparatively low and level.

CLIMATE AND PLANT LIFE

With the possible exception of California, New Mexico has the most varied climate in the United States. This is due to the fact that the highest country is in the northern and the lowest in the southern part of the State.

Near the Colorado State line, where altitudes over 10,000 feet above sea level are common, the snow fall is heavy and the summers are short and cool. Many of the high peaks of this region are above timberline. The annual precipitation is sufficient to sustain profuse growth of evergreens and aspens. At altitudes over 10,000 feet the Engleman spruce and the Colorado blue spruce are the most common evergreens while aspen is the most common of the deciduous trees. At altitudes below 10,000 feet the Douglas fir, and the Pondorosa or western yellow pine are more common. At the more moderate altitudes of 5,000 to 8,000 feet, the piñon and juniper are the most common species, while in the low lands below 5,000 feet the only native plant life is mesquite, cactus, yucca and other desert varieties, except along streams, where cottonwood, willows, birch, and sycamore are common.

In sharp contrast to the Alpine climate of the mountainous region of the northern part of the State is that of the agricultural regions in the southern part of the State along the Rio Grande valley. Here it seldom, if ever, snows, and severe frosts are not common. Many cultivated varieties of fruit are common, and even citrus fruits can be grown in some localities.

MINING DISTRICTS

MOGOLLON AND COONEY DISTRICTS

Situation

The Mogollon mining district is in the western part of Catron County, N. Mex., about 15 miles east of the New Mexico-Arizona State line. Mogollon, with a population of 300, is the largest settlement in the district; it is situated in the center of the greatest mining activity. The village of Glenwood, with a population of 100, is about 10 miles by road west of Mogollon on the main highway from Silver City. Alma, with a population of 128, is about 5 miles north of Glenwood, near the main highway from Silver City to Alpine. At one time it was the largest settlement in the district, but owing to Indian troubles in the early days and changing economic conditions, the population shifted to other centers.

Stock raising is carried on quite extensively in the adjacent regions, and considerable farming has been made possible around Glenwood and Alma by irrigation with water from the San Francisco River.

Transportation and Power Facilities

Silver City is the nearest railroad station, 47 miles from Glenwood and 57 miles from Mogollon. An improved road from Alma to Cooney Canon serves the mines in the northern part of the district. The road from Glenwood to Silver City and that from Glenwood to Mogollon are graveled and in good condition; the latter having been improved recently by workers of the Civilian Conservation Corps. The Cooney Canon road has not been improved for a great many years and is impassable during a considerable part of the summer rainy season.

Ore and concentrate shipments usually are made to the smelter at El Paso, Tex. Occasional shipments are made to Douglas, Ariz. Silver City, through which nearly all rail shipments are handled, is 175 miles by rail from El Paso and 225 miles by rail from Douglas.

Trucking rates to Silver City are 20 cents per hundredweight for concentrates and 30 cents per hundredweight for back freight. The freight rate from Silver City to El Paso is $1.15 a ton on minimum cars of 25 tons, with no sliding scale.

There are no power facilities in the district. All of the mines generate their own power with Diesel or gasoline generating plants. Fuel oil costs around 6-1/2 cents a gallon in Mogollon. The cost of generating current is about 1.5 cents per kilowatt-hour at the larger plants.

History and Production Statistics[3]

The first discoveries in the district were made by Sergeant Cooney of the Eighth U. S. Cavalry, who observed the prominent vein outcrops while making scouting expeditions into the region. In the fall of 1875, when his enlistment expired, he went back into the district and located claims that were later developed into the Cooney mine.

The first ore was shipped in 1879, but in 1880 the valley settlements were nearly wiped out by an attack by Victorio and his Apaches, and Cooney was killed while assisting in the defense of the settlement of Alma. With the repulse of Geronimo and his raiders in 1885 the Indians ceased to be a menace and mining development progressed rapidly.

It has been estimated that prior to 1905 the production was about $5,000,000 in gold, silver, and copper. The first veins developed were those that outcrop in the valley of Mineral Creek; efforts were first confined to rich oxidized pockets near the surface. Later on, ore shoots were developed and mined in the lower silver-bearing sulphide zone in the Maud S and the Deep Down mines in the canyon of Silver Creek below Mogollon. The development of the cyanide process aided greatly in the exploitation of the low-grade ore bodies. As the richer surface ores became exhausted, many of the smaller properties were consolidated. Today the major portion of the producing properties of the district are owned, leased, or held under option by the Mogollon Consolidated Mines Co., and the Black Hawk Consolidated Mining Co.

Table 4 shows the production of the Mogollon district in gold, silver, copper, and lead from 1904 until 1935.

Topography

The elevation of the San Francisco River valley at Alma is about 4,900 feet. Between the river valley and the Mogollon Mountains is a series of flat-topped mesas ranging in elevation from 5,400 to 6,000 feet. Just east of these mesas the Mogollon Mountains rise steeply to an elevation of 7,500 feet. Deep box canyons have been cut in these mountains by all the principal streams.

---

[3] Ferguson, Henry G., Geology and Ore Deposits of the Mogollon Mining District, New Mexico: Geol. Survey Bull. 787, 1927.

TABLE 4. – Gold and silver production of the Mogollon District,
New Mexico, 1904 to 1935.[1]

Year	Ore, tons	Gold	Silver, fine ounces	Copper, pounds	Lead, pounds	Total value
1904	11,276	$ 61,880	79,014	422,308	—	$162,484
1905	15,534	97,158	240,934	295,175	—	288,735
1906	16,075	127,907	268,576	—	—	307,847
1907	20,698	105,431	418,338	150,000	—	411,516
1908	19,546	116,418	278,939	—	—	264,256
1909	23,945	111,464	249,413	46	49	241,167
1910	50,515	304,210	595,669	—	—	625,871
1911	102,219	531,358	1,067,038	1,873	1,862	1,097,206
1912	101,361	524,858	1,093,158	184	386	1,197,197
1913	115,739	619,886	1,306,766	4,418	1,217	1,409,912
1914	136,124	629,102	1,410,327	—	590	1,409,035
1915	119,716	509,165	1,301,059	—	2,426	1,168,916
1916	118,257	373,068	1,008,483	858	3,232	1,037,084
1917	111,934	258,620	722,642	414	1,593	854,327
1918	56,540	119,710	302,902	235	283	422,690
1919	56,531	148,136	382,800	9,521	—	578,643
1920	41,895	125,631	329,489	582	2,050	485,045
1921	48,870	126,791	310,774	—	—	436,565
1922	48,106	142,123	322,460	—	—	464,593
1923	47,644	179,351	398,714	—	—	506,296
1924	72,736	230,960	618,094	—	—	645,083
1925	51,118	163,522	449,659	—	—	475,613
1926	976	12,444	28,404	24,400	—	33,584
1927	89	4,678	3,591	—	—	6,714
1928	81	4,189	16,800	17,000	—	16,465
1929	438	15,287	40,437	15,187	—	39,513
1930	6,554	19,270	61,621	2,400	1,000	43,356
1931	6,274	14,967	38,800	1,100	—	26,319
1932	25,228	66,441	136,869	1,000	1,400	105,113
1933	32,914	54,817	126,020	3,000	1,400	99,131
1934	41,736	111,790	121,357	4,300	1,000	190,624
1935	59,637	243,116	274,172	1,000	1,000	440,300
Total	1,560,306	6,153,748	14,003,319	955,001	17,667	15,491,200

1/ Mineral Resources and Minerals Yearbooks, 1904 to 1935.

Geology[4]

The rough country along the west edge of the Mogollon Mountains is due to faulting along the west front of the range. Two distinct periods of faulting took place, both with relative downthrows to the west. The first faulting was along the east side of the San Francisco River valley and the second was two or three miles farther east. The steep cliffs along the west side of the range are the result of the second period of faulting, which took place along a single fault plane.

During the interval between faulting there was great volcanic activity. Great quantities of sandstone and other rocks of sedimentary origin probably represent periods of quiet between volcanic outbursts. The volcanic rocks were later cut by faults, which became filled with quartz and calcite and formed the principal ore deposits of the Mogollon area. The volcanic rocks were eroded and covered with a considerable thickness of gravel, after which the second period of faulting took place. Following this second period of faulting, the gravel was eroded from the higher eastern blocks exposing the volcanic rocks, and deep canyons were cut in the cliffs that form the fault escarpment. The high mesas, consisting of loosely consolidated gravel, between the San Francisco River and the Mogollon Mountains, represent the period of erosion that took place between volcanic activity and the second faulting.

The age of the volcanic eruptions is not known definitely, but some authorities consider that volcanic action took place in the early Tertiary. As there are considerable erosional unconformities, the action must have taken place over a considerable period of time.

The volcanic rocks of the district consist of andesites, latites, rhyolites, and rhyolite tuffs. The thickness of the series of Tertiary and Pleistocene rocks, as figured from the maximum observed thickness of each formation, is over 8,000 feet, over 80 percent being igneous rocks of volcanic origin. The mineral deposits are entirely in the volcanic rocks of the Tertiary period. The rocks of the Pleistocene are loosely consolidated gravels and sedimentary rocks.

The mineral-bearing veins of the district are fillings in fault fissures, the gangue mineral being mostly quartz and calcite. There are two systems of veins -- the older system bearing approximately east and west and the more recent system bearing nearly north and south. The Queen vein, of the latter system, is the largest, most persistent, and most prominent vein in the district. It can be traced for several miles and traverses the entire district from north to south. It intersects many of the important veins of the east-west system, sometimes forming important ore shoots at the junctions. In general, the higher-grade ore has been found in the east west system of veins, but ore shoots of the Queen vein have, as a rule, proved to be wider, more regular, and more persistent.

---

[4] Ferguson, work cited.

The successive layers of igneous rocks, being cut by the two vein systems and subsequently eroded, have resulted in a seemingly impossible jumble of exposed formations. The veins, for the most part, are on the contacts between different rocks, but occasionally they are in one formation.

Water Supply

The lack of a dependable water supply for milling and domestic purposes has tended to impede the development of the district. Silver Creek is the principal source of supply for the Mogollon section and Mineral Creek for the Cooney section. During the dry season, these creeks furnish little or no water and the mills are operated on water from the mines and from wells. Curtailment of production and sometimes complete shutdowns are common during the summer months.

Shallow wells in the bed of the creek are the principal source of supply for Mogollon during the greater part of the year, and during the dry season they are the only source of supply.

The water supply for the limited population of Cooney is controlled by the only operating mine, which makes it possible for them to prevent pollution of the water obtained from wells during the dry season.

Whitewater Creek is about 2-1/2 miles south of Mogollon. Measurements taken over a period of years show that the flow has never been less than one cubic foot per second. It is from 800 to 1,000 feet below the town of Mogollon and, hence, its use as a source of domestic supply would involve the use of expensive pumping equipment. Its use for milling purposes would involve either pumping to the mills at Mogollon or transporting the ore to a millsite on the creek by aerial tram or truck or through a haulage tunnel.

Mogollon Consolidated Mines Co.

General

The Mogollon Consolidated Mining Co., A. F. Barrett, president and treasurer, Fort Worth, Tex., has lease and options on the Last Chance, Deadwood, and Eberle groups, about half a mile southwest of the town of Mogollon. The claims of the Last Chance group consist of the Frieda, Anna E., Last Chance, Top, Gold Dollar, Gold Dollar No. 2, Little Chance, Hub, and Boise City. The Deadwood group consists of the Deadwood and Simburst and the Eberle group of the Eberle and Braton.

The company began operations in July 1935. In July 1936, the production was from the Deadwood and Last Chance claims. The Eberle was temporarily shut down because of curtailed milling operations. The Deadwood and the Last Chance, on the Last Chance vein, adjoin each other where the vein bends to the south. The Eberle is on the Queen vein near its intersection with the Maud S. The Deadwood and the Last Chance mines are operated as a unit through the Deadwood shaft. The Eberle is operated from an adit tunnel, the portal of which is about a half mile north of the Deadwood shaft. The ore is hauled by truck from bins at the portal of the adit tunnel to the Deadwood mill.

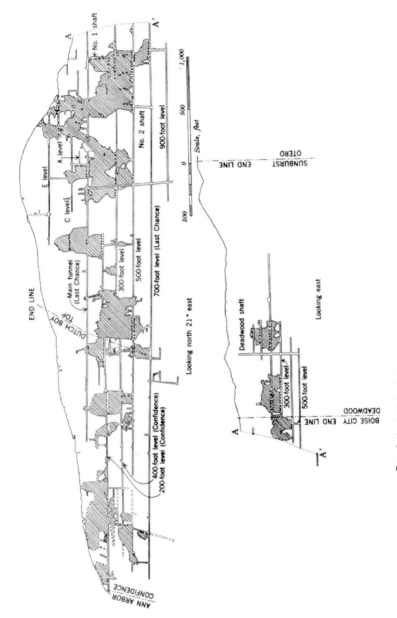

Figure 3.—Longitudinal projection of workings in Deadwood and Last Chance mines, Mogollon, N. Mex.

Geology

The Last Chance vein dips to the northeast at 65° to 75°; it ranges in width from 8 to 12 feet. The gangue minerals are quartz and calcite filling in andesite breccia. The better values are in the footwall quartz. The principal ore minerals are pyrite, argentite, sphalerite, and galena.

Development

The Deadwood mine (fig. 3) is developed through a vertical two-compartment shaft 500 feet deep with level intervals of 75 to 100 feet. There is a total of 3,500 to 4,000 feet of drifting on the Last Chance vein with 500 to 600 feet of crosscutting. The shaft was sunk from the 400-foot level to the 500-foot level, and the fourth and fifth levels connected with the workings of the Last Chance mine by the present operators.

Two large ore bodies were developed and mined above the third level. The ore was milled in a stamp mill which was recently converted to flotation and is being used by the present operators. The Queen vein, which is parallel to this part of the Last Chance vein, was crosscut in several places and some low-grade ore encountered.

The ore bodies did not reach the surface and very little oxidized ore was mined. The development work is on the Deadwood claim, which covers the west end of the Last Chance vein.

The Last Chance mine (fig. 3) is developed through an adit connected with the other levels by raises and winzes. There are eight levels in all, with total drifting on the Last Chance vein in excess of 20,000 feet. The development covers the entire length of the Last Chance and Top claims. The levels are at approximately 200-foot intervals, the lowest being at a depth of about 1,000 feet below the adit and 1,500 feet below the highest part of the outcrop.

No ore has been found on the 900-foot level and only a limited amount on the 700-foot level.

The Eberle mine is developed through an adit about 1,500 feet long on the Queen vein. The portal is near the intersection of the Queen and Maud S veins. The intersection has been exposed in the workings, but development has been confined chiefly to the Queen vein. A short crosscut intersects the Maud S vein, which at this location nearly parallels the Queen vein.

A 50-foot shaft near the adit portal is in ore and several stopes have been opened by former operators.

Mining Methods

Stoping. - When operations were first resumed in 1935, considerable ore was mined from pillars and from the sides of the old stopes where it had been left by former operators. This and the cleaning up and general rehabilitation work incident to starting up an old mine had kept mining costs relatively high. In June 1936, most of the production of 80 tons a day was from one stop on the fourth level on a vein that is parallel to and on the hanging-wall side of a worked portion of the main vein. This stope was from 8 to 12 feet wide, with a rib of waste from 5 to 15 feet thick between it and the old stope. Careful mining was required to keep from breaking through to the old stope in places where the two were close together.

Stoping is by the shrinkage method. The chutes are placed from 20 to 25 feet apart with pillars left in between. The thickness of the pillars, from the back of the drift to the sill of the stope, range from 8 to 10 feet. Raises or crib manways are carried up with the stopes at intervals of 75 to 100 feet, depending on local conditions.

The ore in the stopes is broken by breast rounds drilled with mounted drifting machines. By keeping the breast cuts down to 10 feet or less, most of the ore breaks small enough to be handled in chutes that are about 3 feet wide. Any large pieces that are exposed are blasted before they have a chance to get into the chutes.

The vein bulges and narrows in comparatively short distances along the strike, making it necessary occasionally to make extra set-ups for slabbing ore that was left in the walls unintentionally.

The number of holes required to break a round in the stopes ranges from 10 or 12 in ground that is mostly calcite to 18 or 20 in ground that is mostly quartz. In calcite one set of steel can be used to drill from two to four holes, but in the hardest quartz a set may be required for every hole.

Wet stoper machines are used in the raises and for cutting out chute pockets. Machine-sharpened steel is used in all machines. Detachable bits were tried when the mine was started up but were claimed to be not very satisfactory. One and one-fourth-inch hollow round steel is used in the drifting machines, and 1-inch quarter octagon in the stoper machines.

The machines in the stopes and drifts are operated by one man. The nippers deliver sharp steel to the machines, help set up, and collect the dull steel at the end of the shift.

Tramming and hoisting. - Ore and waste are trammed by hand and hoisted through the Deadwood shaft. A 2-ton skip is used in one compartment of the shaft for hoisting ore. The other compartment is provided with a cage and is used for men and material. A battery locomotive at the portal of the Last Chance adit will be used when operations are resumed in that part of the property.

Sampling. – All sampling is under the direct supervision of the mine engineer. The faces of the stopes and the development headings are carefully sampled each day by channel and grab samples. Enough car samples are taken from the chute and development headings to afford a close check on the mill heads.

Pumping. – Water from the Deadwood workings is pumped through the Deadwood shaft from the 500 or bottom level by a No. 5 Cameron pump. A boiler feed pump on the third level furnishes water for the rock drills. Some of the water from the Last Chance workings is pumped from the portal of the Last Chance adit to the mill and some sinks into the Last Chance shaft, which is in a porous part of the vein.

Ventilation. – Natural ventilation is provided by a large number of connections between levels and by openings left by old worked-out stopes. Forced ventilation is unnecessary.

Labor. – The mine is operated two shifts a day. Most of the common laborers are Mexicans. The foreman, shift bosses, and skilled laborers are Americans.

Classification of labor, number employed, and wage rates, Mogollon Consolidated Mines Co.

Class of Labor	No.	Rate per hour	Total amount for 8-hour day
Miners	13	$0.44	$45.76
Muckers	10	.33	26.40
Timbermen	4	.44	14.08
Pipemen	2	.44	7.04
Pipemen helpers	2	.33	5.28
Trammers	7	.33	18.48
Chuck tenders	6	.33	15.84
Shift bosses	2	5.00 (per day)	10.00
Hoistmen	2	.44	7.04
Cagers	2	.44	7.04
Timbermen helpers	4	.33	10.56
Nippers	3	.38-1/2	9.24
Powder men	1	.50	4.00
Sanitary man	1	.33	2.64
Crusher men	2	.33	5.28
Surface men	1	.38-1/2	3.08
Total per day	62		191.76

Besides the underground force, five men in the blacksmith shop are charged entirely to the mine. The blacksmith shop is run as a separate unit. The payroll is as follows:

```
2 blacksmiths at 44 cents per hour.......... $7.04
2 blacksmith helpers at 33 cents per hour...  5.28
1 drill doctor at 44 cents per hour.........  3.52
Total.....................................   15.84
Total, mine and blacksmith shop............207.60
```

Costs. — The following table shows ore breakage, development footage, drill shifts on stoping and development, and the amount of ore broken and milled for the first half of May 1936:

```
Ore broken in stopes, dry tons.............. 3,713
Ore broken in development, dry tons.........   325
Ore milled.................................. 2,070
Total number drill shifts, stoping.........   477
Total number drill shifts, development......    53
Ore broken in stoping per drill shift, tons.  7.78
Drifting, feet..............................   114
Crosscutting, feet..........................    16
Total development, feet.....................   130
```

Development and ore-breaking costs based upon a total breakage of 4,038 tons and a development footage of 130 feet are as follows:

	Labor	Supplies	Power	Total	Per ton	Per foot
Development:						
Drills and drilling.....	$356.60	$ 34.76	$42.33	$433.69	$0.11	$ 3.33
Tracks and pipe lines...	66.19	2.83	—	69.02	.02	.53
Explosives.............	—	262.80	—	262.80	.07	2.01
Timbering..............	144.04	30.00	1.20	175.24	.04	1.34
Tramming..............	218.29	—	—	218.29	.06	1.67
Hoisting..............	37.67	.96	5.21	43.84	.01	.33
Tools and sundry supplies	—	15.04	—	15.04	—	.12
Tool sharpening........	30.02	8.72	3.78	42.52	.01	.33
Assaying, sampling, and						
surveying.............	22.58	12.66	1.15	36.39	.01	.28
Foreman and shift bosses	50.00	—	—	50.00	.01	.38
Total development....	925.39	367.77	53.67	1,346.83	.34	10.32
Ore breaking:						
Drills and drilling.....	2,455.89	306.53	254.00	3,016.42	.75	
Tracks and pipe lines...	258.25	9.00	—	267.25	.07	
Explosives.............	—	1,200.42	—	1,200.42	.30	
Timbering..............	484.02	101.06	5.03	590.11	.15	
Tramming..............	1,199.86	52.59	15.28	1,267.73	.31	
Hoisting..............	339.10	8.68	46.98	394.76	.10	
Tool sharpening........	265.33	76.94	37.87	380.14	.09	
Tools and sundry supplies	2.60	138.17	—	140.77	.04	
Assaying, sampling, and						
surveying.............	145.89	38.00	3.54	187.43	.05	
Foremen and shift bosses	175.00	—	—	175.00	.04	
Surface tramming........	92.20	—	—	92.20	.02	
Total, ore breaking..	5,418.14	1,931.39	362.70	7,712.23	1.92	
Total, Last Chance...	6,343.53	2,299.16	416.37	9,059.06	2.26	

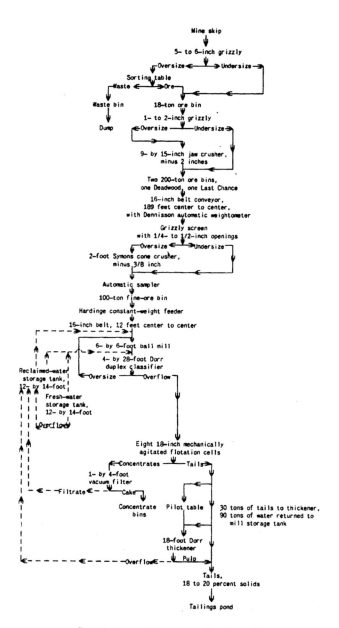

Figure 4.– Flow sheet of the Mogollon Consolidated mill.

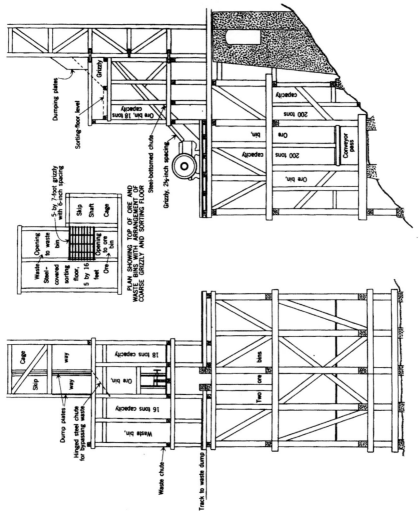

Figure 5.—Ore bins and sorting floor of the Mogollon Consolidated Mines Co., Mogollon, N. Mex.

Dumping plates

Sorting-floor level

Grizzly

Ore bin, 18 tons capacity

Steel-bottomed chute

Grizzly, 2½-inch spacing

Conveyor pass

Ore bin, 200 tons capacity

Ore bin, 200 tons capacity

5- by 7-foot grizzly with 6-inch spacing

Skip   Shaft   Cage

Waste   Opening to waste bin   Steel-covered sorting floor, 5 by 16 feet   Opening to ore bin   Ore

PLAN SHOWING TOP OF ORE AND WASTE BINS WITH ARRANGEMENT OF COARSE GRIZZLY AND SORTING FLOOR

Cage   Skip

way   way

Dump plates

Hinged steel chute for bypassing waste

Waste chute

Track to waste dump

Ore bin, 18 tons capacity

Ore bin, 16 tons capacity

Waste bin,

Two ore bins

Milling Methods

General. – The old Deadwood stamp-mill building of timber construction was utilized for the present mill. The ore is concentrated by flotation. The mill flow sheet is shown in figure 4. According to the agreement under which the building was leased, none of the old equipment could be removed. As a result of this, the arrangement of the new equipment is not as satisfactory as the mill superintendent would like. None of the old equipment is being utilized. The capacity is about 160 tons daily. During June and July 1936, it was operated only three days a week owing to an inadequate water supply.

Sorting. – The ore is dumped from the mine skip on a 5- by 7-foot grizzly that slopes at about 45° and has 6-inch openings. The undersize goes to an 18-ton receiving bin and the oversize to a 5- by 16-foot sorting table (fig. 5) covered with sheet steel. The large pieces of ore are broken to sizes that can be handled by a 9- by 15-inch Blake crusher before they are thrown into the ore bin. The waste is sorted out and thrown into a waste bin. Waste from the mine is deflected into the waste bin by a hinged chute.

Crushing. – From the ore receiving bin the ore goes over a grizzly with a spacing of 1 to 2 inches and a slope of about 45°. The oversize goes through a 9- by 16-inch jaw crusher. The discharge from the jaw crusher and the undersize from the grizzly go into one of two bins of 200-ton capacity each. One bin is for the Deadwood ore and one for Last Chance ore, as the two must be milled separately.

An automatic apron feeder discharges the ore from the 200-ton ore bins onto a 16-inch conveyor belt 189 feet long and equipped with an automatic weightometer. The conveyor belt discharges onto a grizzly screen with 1/4 to 1/2-inch openings. The oversize from the grizzly screen goes through a 2-foot Symons cone crusher that crushes to minus 3/8 inch. The undersize from the grizzly and the discharge from the Symons cone crusher go through an automatic sampler and then to a fine-ore bin of 100-ton capacity.

Grinding. – The ore is taken from the fine-ore bin by a Hardinge constant-weight feeder and is discharged onto a 16-inch conveyor belt 12 feet long. Weightometer tonnage is checked by taking a 6-foot weighed sample from this belt at stated intervals. The tonnage is calculated from the weight of the sample and the speed of the belt as determined by a revolution counter on the Hardinge constant-weight feeder. The conveyor belt discharges into a 6- by 6-foot ball mill driven by a 125-horsepower motor. Cast-iron liners and 3-inch cast-iron balls are used in the ball mill. The ball consumption is about 3 pounds per ton of ore ground. The liners last from 9 to 10 months.

A 4- by 28-foot Dorr duplex classifier is operated in closed circuit with the ball mill. The classifier overflow runs from 60 to 75 percent minus 150 mesh, depending on the mill tonnage. The critical stage of grinding is about 80 percent minus 150 mesh but it has been found difficult to maintain grinding to this degree of fineness at the normal rate of 150 to 160 tons daily with the present grinding and classifying equipment. Reclaimed water is added at the ball-mill intake and reclaimed water and fresh water at the head of the classifier.

Flotation. - The classifier overflow goes to the first of eight 18-inch mechanically agitated flotation cells. Finished concentrates are taken from the first cell. The overflow from the second and third cells is returned to the first and the overflow from the last five is returned to the second. Pulp samples of flotation heads and tails are taken automatically.

Soda ash and coal-tar creosote are added at the feed end of the ball mill at the rate of 1 pound per ton of ore and 0.1 pound per ton of ore, respectively. Xanthate (Z-3) is added at the classifier overflow and at the head of the flotation circuit. Pine oil is added at the head of the flotation circuit.

The value of the mill heads ranges from $7 to $9 a ton, the average being around $8.25 a ton. About 60 percent of the values are in silver and 40 percent in gold. The ratio of concentration ranges from 70: 1 to 80: 1 and the value of the concentrates from $350 to $500 a ton. Tailings loss varies from $1.25 to $1.50 a ton. The following table shows the value of the heads, concentrates, and tails for the first half of May 1936. The values are based on smelter settlement prices of gold and silver, as follows:

Gold, $33.00 per ounce.
Silver, 95 percent of 77 cents, or 73 cents per ounce.

	Tons	Gold, ounces	Silver, ounces	Gold, value	Silver, value	Total	Per ton
Heads........	1,913.65	196.05	12,933.48	$6,502.36	$94,441.44	$15,943.80	$8.33
Tails........	---	36.99	2,034.36	1,226.84	1,485.08	2,711.92	1.42
Heads (less tails)	---	159.06	10,899.12	5,275.52	7,956.35	13,231.87	6.91
Concentrates.	23.196	149.85	9,203.81	4,970.05	6,718.78	11,688.84	6.10

Ratio of concentration = 82.50 to 1 using $\frac{1,913.65}{23.196}$

Mill recovery:

Gold.....81.13
Silver...84.27
Total....82.99

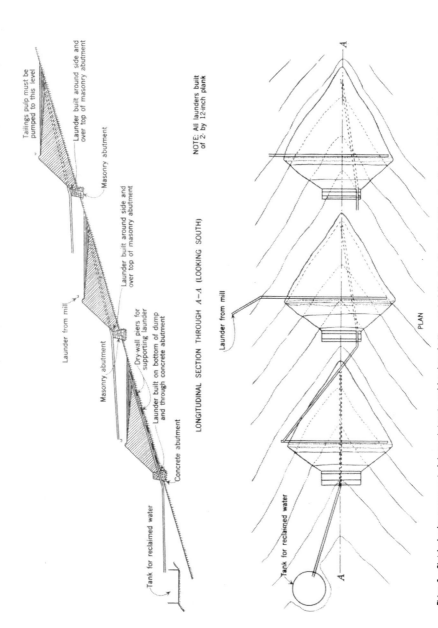

Tailings pulp must be pumped to this level

Launder built around side and over top of masonry abutment

Masonry abutment

Launder from mill

Launder built around side and over top of masonry abutment

Masonry abutment

Drywall piers for supporting launder

Launder built on bottom of dump and through concrete abutment

Concrete abutment

Tank for reclaimed water

NOTE: All launders built of 2- by 12-inch plank

LONGITUDINAL SECTION THROUGH A–A (LOOKING SOUTH)

Launder from mill

Tank for reclaimed water

A

A

PLAN

Figure 6.– Sketch showing arrangement of dams, abutments, and launders of tailings pond of Mogollon Consolidated Mines Co., Mogollon, N. Mex. (Not to scale.)

Dewatering and filtering. -- The concentrate goes through a 1- by 4-foot vacuum filter and then to the concentrate bins. The filtrate from the concentrate is pumped back to the mill storage by a vacuum-seal air lift. The tailing from the flotation cells is divided, one part going to an 18-foot Dorr thickener and the other direct to the tailing pond. A small cut of the tailing goes over an 18- by 48-inch pilot table and then to the Dorr thickener. The overflow from the thickener is pumped to the mill storage tank by a centrifugal pump operated by a 2-horsepower motor. The pulp from the thickener is raised by a diaphram pump and joined with the part of the tailing that goes direct to the tailing pond. The final tailing runs from 18 to 20 percent solids. About 90 tons of water are reclaimed daily from the thickener.

Power. -- Current is generated and distributed at 2,400 volts and stepped down to 440 and 220 for power and lighting. Total connected horsepower in the mill is as follows:

Jaw crusher	15
Automatic apron feeder	1
189-foot belt conveyor	10
Symons cone crusher	25
Automatic ore sampler	1
Hardinge constant-weight feeder	0.25
12-foot belt conveyor	1
Ball mill	125
Dorr classifier	1
Flotation cells	20
Vacuum filter	.50
Vacuum pump	3
Pilot table	.25
Dorr thickener and diaphram pump	3
Centrifugal pump	2
Total	208.00

Tailing disposal. -- The tailing, containing 18 to 20 percent solids is run to the tailing pond, a distance of 2,500 feet, in a wooden launder 8 inches wide and 8 inches deep with a grade of about 5 percent.

The dump is in a natural drainage channel on the side of a hill with a slope of 10° to 15°. Drainage from above is sufficiently controlled to prevent the impounded tailing from being washed away during cloudbursts.

The tailing is deposited on three tiers or benches. Figure 6 shows the arrangement of the launders with relation to the three benches. The upper bench was not in use in June because pumping is required to raise the pulp from the end of the launder to that level.

Past experience seems to have been that there is little danger of the deposits breaking away from the tops of the benches during cloudbursts, but that unless proper provisions are made, movement will start at the bottom, causing the whole mass to move downhill.

Such movements are prevented by concrete or masonry abutments built into the bedrock at the toe of each bench. The tailing is deposited by a launder built at the edge of each bench and sloped up from the top of the abutment at about the angle of repose for the material when dry.

The coarse material settles at the edge of the benches and is used for building up the dam to impound the water containing the slimes. The water, containing some slime, is run through a launder to the bench below and the operations are repeated. From the second bench the water comes off practically clear and is run to a storage tank, from which it is pumped back to the mill.

The launders from the two upper tiers are built around the sides and over the top of the abutment. The one from the bottom tier to the storage tank is on bedrock in the lowest part of the drainage channel, where it is covered by the tailing as it builds up, except at the intake. Additions are made at the intake of the launders as the necessity arises.

Four men are required at the tailing pond, building up the dams and building additions to the launders. The launders must be built on solid rock or on rock piers placed at intervals of 4 or 5 feet. If they are built on top of the tailing and then buried they will be broken and crushed as the material builds up and later settles. The cost of tailing disposal in May was 20 cents a ton.

Labor. -- The mill is operated three shifts a day by 16 men, including the mill superintendent. The payroll is as follows:

1 mill superintendent	Salary.
3 operators	55 cents per hour.
3 helpers	38-1/2 cents per hour.
3 crushermen on Symons cone crusher	38-1/2 cents per hour.
2 laborers at drying and sacking concentrates	44 cents per hour.
4 laborers at the tailings pond	44 cents per hour.

The operation of the jaw crusher at the mine shaft is charged to mine-operating expenses.

Costs. – The following table shows the milling costs for May 1936 (total tons milled, 4,027):

	Labor	Supplies	Power	Total	Per ton
Coarse crushing.....................	$ 731.01	$ 131.73	$ 66.29	$ 929.43	$0.23
Conveying and fine crushing........	339.81	138.29	102.81	580.91	.14
Fine grinding and classifying......	271.25	18.33	653.66	943.24	.23
Balls and liners...................	—	537.07	—	537.07	.13
Flotation..........................	264.13	18.35	110.10	374.58	.09
Reagents...........................	—	665.09	—	665.09	.17
Filtering and dewatering...........	255.11	44.50	47.12	346.73	.09
Concentrate drying.................	212.19	137.04	22.95	372.18	.09
Tailing disposal...................	607.01	168.97	15.28	791.86	.20
Concentrate freight................	—	247.88	—	247.88	.06
Water lines and pumping fresh water	151.95	23.46	35.47	210.88	.05
Assaying and sampling..............	114.55	53.43	4.61	172.57	.04
Supervision........................	250.00	—	—	250.00	.06
Lighting...........................	5.51	35.26	14.14	54.91	.02
Building...........................	27.61	—	—	27.61	.01
Totals.....................	3,230.13	2,219.40	1,072.43	6,504.94	1.61

Freight and Smelter Costs

The concentrates are trucked 57 miles to Silver City by contract truckers. There is generally back freight, such as fuel oil and other supplies. The rate from Mogollon to Silver City is 20 cents per hundredweight. The rate on back freight is 30 cents per hundredweight.

The freight rate from Silver City to the El Paso smelter is $1.15 per ton, minimum car 25 tons, no sliding scale. The total freight from Mogollon to the El Paso smelter is $5.15 per ton.

The smelter pays for metals according to the following schedule:

Gold – $33.166 per ounce.
Silver – 77 cents per ounce for 95 percent if under 500 ounces per ton; 77 cents per ounce for 97-1/2 percent if over 500 ounces per ton.
Copper – Deduct 0.4 units and pay for 95 percent of the balance at market price less 2.525 cents per pound.

Charge and penalties:
    Base – $4.20, no sliding scale.
    Silica – No penalty.
    Iron – No bonus.
    Zinc – 5 percent free; charge $0.30 per unit excess.

Power House and Compressor Plant

The power house is about 1/2 mile south of the Deadwood mine. The building and some of the equipment are part of the old plant used by former operators. The main generating units used by the present operators were installed recently.

The generating equipment consists of two 6-cylinder, 300-r.p.m., 420-horsepower, full Diesel engines, each direct-connected to a 248-kilowatt, 3-phase, 60-cycle, 2,400-volt, 85.5 amperes per terminal generator, and one 2-cylinder, 180-horsepower, 200-r.p.m., semi-Diesel engine direct-connected to a 150-kv.-a., 250-volt, 360.8-ampere, 3-phase generator.

The semi-Diesel unit is used for emergency purposes only. It is connected with transformers at the power house so that the voltage can be stepped up for transmission. Power is transmitted at 2,300 to 2,400 volts.

The compressor plant is in the power house and is taken care of by the power-plant operators. The main unit consists of a 1,100-cubic-foot compressor belt-connected to a 150-horsepower, 2,200-volt, 39-ampere, 60-cycle, 3-phase induction motor. The emergency unit consists of a 300-cubic-foot compressor direct-connected to a 1-cylinder, 90-horsepower, semi-Diesel engine.

The plant is run three shifts a day by a chief engineer or operator and two assistant operators. The chief engineer received $5 a day and the two assistants 49-1/2 cents per hour.

Fuel oil cost 6.8 cents a gallon in Mogollon. The plant uses about 550 gallons a day, or 0.5 to 0.6 pound per horsepower-hour. The cost of generating current is about 1-1/2 cents per kw.-hr.

The power costs per ton of ore milled in May were as follows:

	Amount	Tons ore milled	Cost per ton
Mining	$362.70	4,027	$0.090
Development	53.67	4,027	.013
Milling	1,120.78	4,027	.278
Total	1,537.15		.381

Black Hawk Consolidated Mining Co.

General

The Black Hawk Consolidated Mining Co., Theodore Trecker, Milwaukee, Wis., president, has a lease on 40 patented claims about 1/2 mile northwest of the town of Mogollon. The claims belong to the Lehigh Metals Co., Eugene Schimff, of Scranton, Pa., president. The property includes the Little Fanney mine and 200-ton cyanidation mill and Diesel power plant, and a number of other groups of claims to the north of Silver Creek.

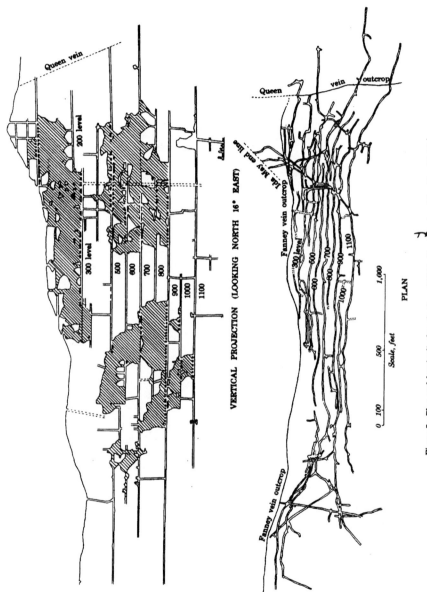

VERTICAL PROJECTION (LOOKING NORTH 16° EAST)

Queen vein

200 level

300 level

500 600 700 800

900 1000 1100

Queen vein outcrop

Fanney vein outcrop

Fanney vein outcrop

300 level

500 600 700 800 900 1000 1100

Little May end line

PLAN

Scale, feet

0 100 500 1,000

Figure 7.—Plan and longitudinal projection of Little Fanney mine, Mogollon, N. Mex.

The Little Fanney mine (fig. 7) is one of the largest mines in the district. It is extensively developed through the Little Fanney shaft, which is 1,100 feet deep. The Little Fanney vein has been mined laterally for a distance of 2,000 to 2,500 feet, and down to the 1,100-foot level. There are several winzes below the 1,100-foot level but very little drifting and crosscutting. From some of the levels, the workings have been extended into the territory of adjacent properties.

The production of the Little Fanney up to 1919 was $4,869,000. The only production since then was in 1923, 1924, and 1925, during which time a limited tonnage was mined from the Little Fanney vein and from adjoining properties through the Little Fanney workings. The mill was operated in 1924 and 1925 but was shut down in January 1926 and remained idle until 1933, when it was partly remodeled for milling ore from another property. The mine has been idle since 1926. The shaft and hoisting equipment are kept in operating condition because of the necessity of pumping water for milling.

Present mining operations are on the Queen vein on the Andrew Jackson Consolidated claim about 1/4 mile northeast of the Little Fanney shaft. Nearly all development on this part of the property has been done since 1932. Development consists of an adit crosscut tunnel about 200 feet long, a single compartment incline shaft 300 feet deep on the Queen vein, and about 600 feet of drifting on each of the four levels.

Mining Methods

Stoping. -- The stoping width of the ore ranges from 10 to 25 feet and averages about 15 feet. The shrinkage method of mining is used exclusively, with the chutes placed at 20- to 25-foot centers. Eight- to 10-foot pillars are left between the back of the drift and the sill of the stope. The ore from these pillars is broken and sent to the mill as soon as a level is mined out. Stoper machines are used in the raises and for cutting chute pockets. In the stopes, breast rounds are drilled with drifter machines. One and one-fourth-inch steel with machine-sharpened bits is used in the drifters.

Drifting. -- The drifts are run 12 feet wide and double tracked. This is to speed the tramming enough for a production of 100 tons or more a day from one or two levels. Two machines are used in a 7- by 12-foot drift heading, each machine drilling from 16 to 20 holes for a round, depending on the hardness of the rock.

In July 1936, the drifts on the 300-foot level were being extended north and south from the shaft. The south drift, which was in quartz, required 20 holes on each side, or 40 holes in all. The north drift, which was in material containing more calcite, required 16 holes on each side, or 32 holes in all.

The machine men were given a clean set-up every morning. Drilling and blasting was done on the day shift and mucking on the graveyard shift. A helper in each heading brought the steel from the shaft to the machines, helped set up, and acted as chuck tender for the two machines after getting started.

Costs and other drifting data were not available.

Haulage and hoisting. — Tramming on the levels is by hand in ton cars. The ore is dumped into skip pockets, from which it is drawn into a 1-ton skip and hoisted to the main adit haulage level and dumped into an ore bin. Haulage from this bin to the mill bins is by a locomotive powered by a gasoline engine. The locomotive pulls a train of nine 1-ton cars. The airline distance to the mill is about 1/4 mile, but to keep on grade the car track is built around a ridge, making the distance between 1/2 and 3/4 mile.

Labor. — The company pays the prevailing wage scale of the camp. Muckers, nippers, and trammers get 33 cents per hour (1936). Timbermen, machinemen, hoistmen, and other skilled laborers get 38-1/2 to 40 cents an hour. The mine is normally operated on a 2-shift basis, but during June and July 1936, when development was being done on the third level, considerable mucking was done on graveyard shift.

From 65 to 70 men are employed underground to produce 100 tons a day. In July 1936 the number of machine shifts was as follows:

Place	No. machines	No. shifts	No. machine shifts
No. 3 level, south drift	2	1	2
No. 3 level, north drift	2	1	2
No. 3 level, stope preparation	2	2	4
No. 2 level, breaking pillars	1	2	2
Total			10

Costs. — From January 1 to May 1, 1936, 12,190 tons of ore were mined and milled. During the early winter, one of the generating units of the power house was shut down for nearly a month for repairs. During this time enough current was generated by the other units for the mill and the mine hoist only. Development and ore breaking were discontinued, but the mill was kept operating on the broken reserve, practically all of which was drawn by the time the necessary repairs were made, so that the compressor could be operated and ore-breaking and development work continued. The breakdown in the power house, together with difficulties that had to be overcome in catching up with the development work, is reflected in relatively high mining costs for the first four months of the year. Mining costs from January 1 to May 1, 1936 are as follows (tons mined and milled, 12,190):

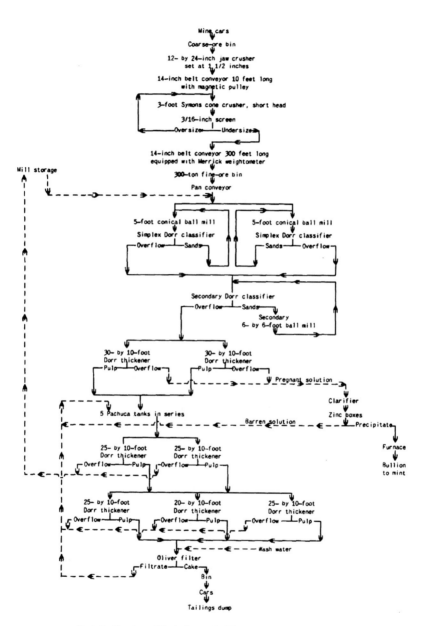

Figure 8.- Flow sheet of Little Fanny mill of Black Hawk Consolidated Mining Co.

	Labor	Supplies	Explosives	Depreciation	Total
Supervision	0.166	—	—	—	0.166
Drill sharpening and steel	.099	0.100	—	—	.199
New rock drills	—	—	—	0.118	.118
Ore breaking	.660	.068	0.319	—	1.047
Drawing reserves	.152	.056	—	—	.208
Mucking and tramming	.442	.005	—	—	.447
Hoisting	.144	.004	—	—	.148
Motor haulage	.265	.023	—	.021	.309
Total	1.928	.256	.319	.139	2.642

Milling Methods

General. — The ore is treated by cyanidation in the Little Fanney mill, which was constructed by the original owners of the Little Fanney mine. A number of changes were made by the present operators. Flotation was tried, but the results were not very satisfactory and the equipment was changed back for cyanidation (fig. 8).

The mill capacity is 200 tons a day, but during the dry season there is not sufficient water for more than 100 to 125 tons a day. The water supply is from the Little Fanney mine shaft and from Silver Creek.

The mill heads run from $9.00 to $12.00 a ton. About 40 percent of the values are gold and 60 percent silver. Mill recovery ranges from about 88 to 93 percent. According to Ira Wright, manager of the company, recovery is better when the mill is not operated at full capacity, due, perhaps, to better agitation effected in the Pachuca tanks.

Crushing. — Ore from the mine cars is dumped, without being sorted, into a coarse-ore bin. From the coarse-ore bin it is fed by gravity through a chute gate to a 12- by 24-inch jaw crusher that crushes to minus 1-1/2 inch. The discharge from the jaw crusher is taken to a 3-foot, short-head, Symons cone crusher by a 14-inch belt conveyor about 10 feet long equipped with a magnetic pulley to remove tramp iron. The cone crusher is in closed circuit with a 3/16-inch screen. A 14-inch belt conveyor 300 feet long, equipped with a Merrick weightometer, carries the ore to an ore bin of 300-ton capacity in the main building.

Grinding. — The ore is conveyed from the fine-ore bin by a pan conveyor to two 5-foot conical mills, each operated in closed circuit with a Simplex Dorr classifier. The overflow from the Dorr classifiers goes to a secondary Dorr classifier, which is operated in closed circuit with the secondary 6- by 6-foot ball mill. The overflow from the secondary classifier is 50 percent minus 200 mesh.

Three-inch balls are used in the primary conical ball mills and 2-inch in the secondary mill. The ball consumption is 3.4 pounds per ton in the primary mills and 1.6 pounds per ton in the secondary mill.

The ends of the ball mills are lined with standard manganese steel end liners and the periferies with 4-inch steel rail sections. The sections are set on end in neat cement and wedged in with steel wedge bars. The cost of lining a conical mill with rail sections is about as follows:

Material, including end liners and labor of cutting
rail section with an acetylene torch and making
wedge bars...................................... $455.00
Labor of installing rail sections and end liners..... 40.00
Total.. 495.00

The cost of lining entirely with standard manganese liners is $1,114. The rail lining lasts about 7 months and the standard manganese liners a little less. From 24 to 36 hours is required to line with rail sections. In order to permit the cement to set properly, the mill should not be used for 12 to 24 hours after it has been lined.

Cyanidation. - The overflow from the secondary classifier is divided into two equal parts. Each part goes to a 30- by 10-foot Dorr thickener. The overflow from these thickeners is the pregnant solution. The pulp goes through five Pachuca tanks in series and then through two secondary thickeners in parallel. The pulp from the secondary thickeners goes through three tertiary thickeners in parallel and the pulp from these goes through an Oliver filter. The overflow from the secondary thickeners is returned to the mill storage tank. The overflow from the tertiary thickeners and the filtrate from the Oliver filters are returned to the Pachuca tanks.

Litharge is added at the feed of the primary ball mills at the rate of 1/3 of a pound per ton of ore, and lime at the rate of 8 pounds per ton of ore. Cyanide is added at the first and third Pachuca tanks. The cyanide solution is kept at a strength of 4.2 pounds per ton at the first tank and 5 pounds per ton at the third tank. The cyanide gives better results if it is completely dissolved before it is added.

Precipitation and melting. - The pregnant solution from the primary thickeners goes through a clarifying tank and then to zinc boxes. The precipitate from the zinc boxes is melted down in a furnace and sent to the Mint. The barren solution from the zinc boxes is returned to the Pachuca tanks.

Tailings disposal. - After the tailings are filtered, they are run to a bin, from which they are drawn into a mine car and trammed to the tailings dump, which is on a steep hillside. The moisture content of the filtered tailings is from 7 to 9 percent. With this amount of moisture they lie at a greater angle of repose than would be the case for thoroughly wet or thoroughly dry tailings. This has resulted in the sides of the dump being steeper than desirable and has been the cause of some trouble in the past. During continued heavy rains the whole mass of material, upon becoming wet, slides down the hill.

Cost of tailing disposal is 23.9 cents per ton of ore milled. Of this, 19.6 cents is for labor and 4.3 cents for material.

Power. -- Power is generated at the company's central Diesel plant. Current is generated and distributed at 440 volts. The motor hookup in the mill is as follows:

Jaw crusher...	50
Symons cone crusher...................................	25
Long conveyor...	10
Pan conveyor..	5
Hardinge mills.......................................	100
Marcy mill...	50
Classifiers..	15
Blower...	25
Seven Dorr thickeners (estimated)....................	40
Oliver filter and vacuum pump (estimated)............	5
	325

Costs. -- Detailed milling costs, January 1 to May 1, 1936, are as follows (tons milled, 12,190):

	Labor	Supplies	Fuel	General	Total
Supervision..........	--	--	--	0.048	0.048
Crushing.............	0.216	0.141	--	--	.357
Grinding1/...........	.106	.219	--	--	.325
Cyaniding2/..........	.156	.666	--	--	.822
Tailings disposal....	.196	.043	--	--	.239
Smelting.............	.034	.038	0.013	--	.085
Defrayed charges.....	--	--	--	.014	.014
Totals...............	.708	1.107	.013	.062	1.890

1/ Includes balls, $0.136 per ton.
2/ Includes cyanide, 0.352 per ton; lime, 0.062 per ton; and litharge, 0.020 per ton.

Power House and Compressor Plant

The power house was built in 1911; the original equipment, with a few minor changes, is still in use. Enough current is generated to supply the needs of the company and to furnish lighting for the town of Mogollon.

All current is generated at 440 volts. One line of 6,600 volts is used for transmitting power to the hoist at the mine. All other power is distributed direct from the generators.

The generating units consist of two 187-kilowatt, 3-phase generators, each direct-connected to a 260-horsepower De La Vergne semi-Diesel engine; one 100-kilowatt, 3-phase generator, belt-connected to a 180-horsepower De La Vergne semi-Diesel engine; and one 180-kilowatt, 3-phase generator belt-connected to a 260-horsepower De La Vergne semi-Diesel engine.

The plant requires from 0.52 to 0.54 pound of fuel per horsepower-hour. The cost of the fuel oil is 6-3/4 cents per gallon at the plant. The output is from 250,000 to 275,000 kw.-hr. per month at a direct cost of 1.545 cents per kw.-hr. Detailed power costs are as follows:

Supervision and labor..........	$0.00063	per kw.-hr.
Labor.........................	.00246	do.
Supplies......................	.00008	do.
Fuel..........................	.00692	do.
Lubrication...................	.00231	do.
Total operation...........	.01240	do.
Depreciation..................	.00304	do.
Total.....................	.01544	do.

The compressor is in the power plant and is taken care of by the plant operators. It has a capacity of 967 cubic feet per minute, altitude rating, and is run by a 125-horsepower electric motor.

Cooney Mining Co. Inc.

Situation and General

The property of the Cooney Mining Co., Inc., Robert W. Lyons, president, Shoreham Bldg., Washington, D. C., is in Cooney Canon about 1-1/2 miles north of the town of Mogollon and 1/2 mile east of the old town of Cooney, now practically deserted. Glenwood, about 10 miles distant, is the mailing address for the mine. There is an unimproved road from the mine to the main highway about 7 miles to the west.

The claims, twelve in all, are owned by John Hoover and Wm. J. Wetherby of Mogollon. The group is operated under bond and lease by the Cooney Mining Co., Inc. The claims include some of the first locations in the district. Patents were issued for some of them as early as 1880, but a number are still held by location. Five of the claims, the Copper Queen, Colonel, Apache, Victorio, and Geronimo, are on the Queen vein; they include the Queen mine, which is the present source of production. The Majestic, Majestic No. 2, Silver Lead, Silver Lead No. 2, Silver Twig, Silver Twig No. 2, and Sun are mostly on the Twig vein. They have been developed only to a limited extent.

The Queen mine was first opened up in 1895. It had been idle for a great many years when present operations began in 1934. Very little information is available with regard to output, but the limited extent of stoping would indicate a small production.

Geology

The vein dips from 65° to 75° to the east. It ranges in width from 3 to 10 feet and averages over 6 feet. The gangue mineral is principally calcite with inclusions of andesite fragments. Quartz occurs sparingly, but, as in the mines farther to the south, it is more abundant in the oreshoots than in other parts of the vein.

The ratio of gold to silver is higher than in most parts of the district, being about 1 to 25 by weight instead of the usual 1 to 50 or less. Copper minerals also occur in greater abundance than in other parts of the district, but not enough to interfere seriously with cyanidation.

The Queen mine was thoroughly sampled by cutting channel samples at 5-foot intervals, and assay plans were made of all workings. The values, as indicated by the samples from a typical block, are as follows:

Sample No.	Gold	Silver	Value
1	0.21	5.00	11.20
2	.32	8.00	17.36
3	.20	5.40	11.16
4	.21	5.00	11.20
5	.316	8.13	17.31
6	.200	6.20	11.77
Average	.243	6.29	13.33

In many places where the vein is narrower than the practical stoping width there is an overbreak and dilution of 20 to 25 percent. Taking this into consideration, the above samples will check closely with the mill heads that run about $10 a ton.

Development

The development work in the Queen mine is entirely on the Queen vein on the north side of Cooney Canon. An adit level, known as the Queen level, is 1,200 feet long and is connected by raises to two levels above, the first 700 feet long and the second 80 feet long. A winze from the Queen level, 120 feet deep, was not accessible. The level interval is 120 feet. The upper levels do not have direct connections to the surface. There are several worked-out stopes on the Queen level, but the amount of virgin ground is relatively large compared to most old mines in the district.

Development and prospecting at the Queen mine in June 1931 consisted of drifting north on the Queen level. The vein, being well defined, is easy to follow and very little crosscutting is done. In the spring of 1936 an adit drift was started on the Queen vein on the opposite side of the canyon from the Queen mine at an elevation of 75 to 100 feet above the Queen level. In July 1936 this drift was in 180 feet from the portal; the last 50 feet was in ore running from $6 to $8 a ton. This is the only development of importance on the Queen vein on the south side of Cooney Canon. One machine, working two shifts a day, was breaking from 20 to 30 tons. The ore was being dumped on a stock pile at the portal until a jigback tram could be built to transport it to the mill.

Mining Methods

Stoping. -- During June and July 1936 the entire output of ore of about 50 tons a day was from development and stope-preparation work in the Queen mine. One machine was being operated two shifts a day on development work and another two shifts a day on stope-preparation work.

The shrinkage method of stoping is to be used. Some of the old worked-out stopes were started by leaving 5- to 8-foot pillars between the back of the drift and the sill of the stope. In other places stoping was started at the back of the drift with stulls and lagging used to form the sill of the stope. The chutes were at intervals of 15 to 20 feet.

Haulage. -- The ore is trammed by hand through the Queen level to a bin just beyond the portal and at the top of the mill. Ore from the upper levels is transferred to the Queen level through raises used as ore passes.

Labor. -- The mine payroll consists of 6 miners at $5 a day, 7 muckers at $4 a day, and 6 nippers and trammers at $4 a day, a total of $82 a day.

Compressor plant. -- The compressor plant consists of two motor-driven compressors, each of 320-cubic-foot capacity and connected with multiple V-velt drives to 50-horsepower motors. A 670-cubic-foot compressor belt-connected to a 75-horsepower motor is used as a stand-by. The plant is in the same building as the power plant and is taken care of by the power-plant operators.

Milling Methods

General. -- An old stamp mill, with alterations and improvements, is used as part of the plant and treats 50 to 60 tons of ore a day by cyanidation. A 4- by 8-foot rod mill and a Dorr duplex classifier were added to the grinding circuit and a Merril-Crowe precipitation plant was substituted for the zinc boxes that were used originally. The flow sheet is shown in figure 9.

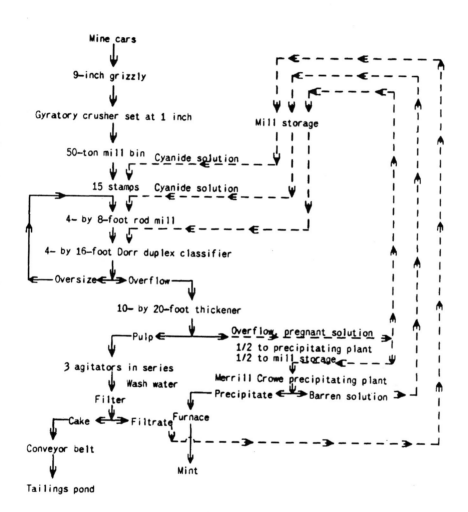

Figure 9.— Flow sheet, Cooney Mining Co. mill.

Crushing. – The ore is dumped from mine cars onto a grizzly with 9-inch spacing that discharges into a coarse-ore bin. The ore is drawn from the coarse-ore bin through a chute gate and into a gyratory crusher that crushes to minus 1 inch. The gyratory crusher discharges directly into a fine-ore bin of about 50 tons capacity.

Grinding. – From the fine-ore bin the ore goes through a battery of 15 stamps and then into the 4- by 8-foot rod mill in closed circuit with a Dorr duplex classifier. Lime is added at the stamp feed at the rate of about 4 pounds per ton of ore. Solution is added at the stamps, rod mill, and classifier.

Cyanidation. – The classifier overflow goes to a 10- by 20-foot thickener. The overflow from this thickener is the pregnant solution and is divided into two equal parts; one part goes to the mill storage tank and the other part to the Merril Crowe precipitation plant. Strong cyanide solution is added to the part that goes to the mill storage tank. The solution at the head of the circuit is kept at a strength of about 4 pounds of cyanide per ton of solution. The pulp from the primary thickener goes through three agitators and then to an Oliver filter. The filtrate from the filter is returned to the mill storage tank and the filter cake is taken by a conveyor belt to the tailings pond.

Precipitation. – The barren solution from the Merril Crowe precipitation plant is returned to the mill storage tank. The precipitate is melted in a furnace and sent to the mint.

The motor hook-up in the mill is as follows:

```
            Crusher.................. 15
            Stamps................... 50
            Rod mill................. 50
            Classifier...............  3
            Thickener................  3
            Agitators................ 12
            Filter...................  5
            Conveyor................. 10
                 Total..............148
```

Costs and metallurgical data were not available. According to the assayer and the mill superintendent, the mill was making about a 90 percent recovery. The heads ran about $7 a ton in gold and $3 in silver.

Power Plant

The power plant, which is in a building adjoining the mill, consists of a 2,300-volt, 250-kilowatt, 3-phase generator direct-connected to a 325-horsepower De La Vergne semi-Diesel engine. A 30-kilowatt, 2,200-volt, 3-phase generator, belt-connected to a 40-horsepower Diesel engine, is used for emergency purposes. Current is generated at 2,300 volts and stepped down to 440 for use in the mill motors.

Four men per day operate the compressor plant and power house. The wages are $4 a day.

Idle Mines

Bearup

The Bearup mine is on the north bank of Silver Creek, 9 miles northeast of Glenwood and 3 miles west of Mogollon. The property, consisting of several unpatented claims, belongs to the Bearup estate, Wm. Bearup, of Mogollon, administrator.

The mine is just west of the range front fault scarp in country that has been very much broken up. The ore bodies consist of high-grade bunches of highly oxidized ore containing very few sulphides. There is very little silver and most of the gold is free. The mine has had a limited production but statistics thereon are not available.

There is a total of 3,000 to 4,000 feet of development work, consisting principally of a haulage level running east and west on the vein and another shorter level 41 feet above this.

The surface equipment consists of a flotation mill of 18 to 20 tons daily capacity, a blacksmith shop, a portable gasoline compressor, a boarding house, and it quarters 10 to 15 men.

The mill was run by a 75-horsepower gasoline engine connected to the mill machinery by belts and a line shaft. The crushing and grinding equipment consist of a jaw crusher and two 3-foot ball mills in closed circuit with a simplex classifier. The flotation circuit consists of three mechanically agitated cells and two pneumatic cells. The mechanically agitated cells are at the head of the circuit. The first made a finished concentrate. The overflow from the other four was returned to the head of the circuit.

The mill was built by a lessee, who operated it for only a short time. In July 1936 the mine and mill were idle.

Maud S.

The Maud S. mine is on the north bank of Silver Creek Canon, about 1/2 mile west of Mogollon. It was in this mine that the first silver-bearing sulphide ore in the district was discovered. The mine has been idle for a great many years and very little is known of the production. It is estimated that it has produced from $600,000 to $800,000 in gold and silver.

Development consists of an adit 1,100 feet long with an incline shaft near the portal and another at the end of the adit. The shaft on the west end reaches a depth of 700 feet on the incline and the one on the east end a depth of 250 feet on the incline. Besides the adit, there has been extensive development on the 500- and 700-foot levels from the west shaft.

The development is all on the Maud S. vein, which strikes roughly N 60° W and dips 60°N.

Deep Down

The Deep Down mine adjoins the Maud S. on the east. It is on the south bank of Silver Creek on a portion of the Maud S. vein that bears nearly north and south. The Maud S. forms a junction with one of the veins of the Queen group at the Eberle mine, where it has been exposed but not explored.

Stoping at the Deep Down mine was entirely from oxidized ores near the surface. The workings have been inaccessible for many years and not much is known about the extent of development or the extent of the production, which, however, is about $75,000. The workings are connected with those of the Maud S. through a shaft on the Maud S property.

Confidence

The Confidence property, owned by the Whitewater Mining and Power Co., consists of the Dutch Boy, Blackbird, and Confidence claims and adjoins the Last Chance property on the west (fig. 3).

The claims are on the Last Chance vein, which has been developed to a depth of about 1,000 feet. The underground workings are connected with the Last Chance workings on several levels. No ore has been mined below the 450-foot level and very little below the 250-foot level.

The mine was last operated in 1925. Production has been about $1,500,000. A mill on Whitewater Creek about 3 miles west of Glenwood was at one time operated in conjunction with the mine. This mill was later destroyed by fire.

Cooney

The Cooney mine is on the south bank of Cooney Canon on the Silver Bar vein at its intersection with the Twig vein. This was one of the first mines located in the district and was at one time the most productive. It was staked by Sergeant Cooney, who discovered the district. The Silver Bar vein strikes nearly north and south at its intersection with the Silver Twig vein, but north of the intersection it strikes northwesterly.

The ore contains chalcopyrite, bornite, chalcocite, and some galena. It runs higher in copper than any other ore in the district. It is stated that some of the ore contained 45 percent copper, but it was mined for the gold and silver. The copper mineralization decreased at depth and was replaced by iron.

The mine is extensively developed by an adit, a winze 760 feet deep, and six levels below the adit level. Most of the workings are caved, and no very definite information is available as to the extent of the ore bodies. According to old descriptions, some of the stopes are 30 feet wide and many are 12 to 15 feet wide.

Production up to 1905 was about $1,000,000, but the mine was operated since then. The total production has been estimated as high as $1,700,000.

Gold Dust

The Gold Dust mine, Theodore Carter, owner, El Paso, Tex., is about 1 mile south of Mogollon; the distance by road, however, is about 4 miles. It is about 1 mile north and 1,500 feet above Whitewater Creek. There are three claims on the Gold Dust vein, which bears approximately east and west, and two on the Queen vein, which bears north and south.

Development consists of two adits on the Gold Dust vein about 350 feet apart vertically. The lower tunnel is about 1,300 feet long and the upper about 500 feet long.

N. O. Bagge, of Glenwood, states that channel samples that he cut in the upper tunnel averaged around $16 a ton in gold and silver, but that the enriched portion of the vein is only 2 to 3 feet wide. The other part of the vein consists of clay on the walls which it would be difficult to keep out of the ore except by hand sorting, and which is very difficult to handle in a mill.

The lower tunnel is caved and inaccessible. Some small shipments of ore were made to the old Deadwood mill, but the production was small. The work was done by hand.

Ann Arbor

The Ann Arbor mine, Harry Herman estate, owner, is on the west extension of the Last Chance-Confidence vein. It is about 1/2 mile from the Confidence mine and 2 miles southwest of Mogollon.

Development consists of about 2,000 feet of adit on the Confidence vein.

Plant and equipment consist of a gasoline-driven compressor at the mine and a mill on Whitewater Creek about 3 miles east of Glenwood.

The mill was operated only for a few days intermittently on a very limited production. Production has been negligible.

Alberta

The Alberta mine, owned by the McKay estate of Pittsburgh, is about 1 mile north of Mogollon. There is one claim on the Queen vein adjoining the Little Fanney property and several on an east-west vein known as the Ida May. All claims are patented. It was operated by the owners from 1910 to 1915. It is developed by a shaft 200 feet deep and several hundred feet of drifting, a large part of which was done through the workings of the Little Fanney mine. No figures on production are available.

Golden Eagle

The Golden Eagle mine, owned by Sill Gamul of Mogollon, is about 1-1/4 miles northwest of Mogollon. It consists of two claims on the Golden Eagle vein, which bears N 30° W and dips 67° E.

There is a shaft from the surface about 70 feet deep and a tunnel 500 feet long on the vein. The shaft is not quite deep enough to connect with the tunnel.

Several small shipments of ore running about 1/2 ounce in gold have been made. The ratio of gold to silver is high – about 1 to 9 by weight. This is probably due to the oxidized condition of the ore near the surface and might point to the possibility of silver enrichment at depth.

Trilby

The Trilby mine, owned by ex-Senator Bursum, of Socorra, is about 1 mile northwest of Mogollon. It is developed by a shaft 200 feet deep and several hundred feet of drifting on the Trilby vein. It is connected with the Little Fanney workings and is accessible only through the Little Fanney mine. Most of the very limited production was shipped to the Little Fanney mill.

HILLSBORO DISTRICT

Situation

The Hillsboro district is in Sierra County, about 15 miles west of the Rio Grande River. The town of Hillsboro, with a population of 403 in 1930, is the county seat and is southwest of the district on the Percha River. It is the only settlement of importance in the immediate vicinity.

Facilities

The district is connected by good graveled highways to Lake Valley 18 miles south, Hot Springs 35 miles northeast, and Santa Rita 35 miles west. Rail shipments are handled through Lake Valley. El Paso, where the nearest smelter is located, is 120 miles from Lake Valley by rail and 126 miles from Hillsboro by highway. Trucking costs to El Paso are around $5 a ton.

The Percha River furnishes a supply of water that could be utilized for milling and domestic purposes, although it dwindles to a mere trickle during the dry season and sometimes dries completely. Most of the domestic needs of the town of Hillsboro are supplied by wells; some old mills in the district utilized water from the mines. A sufficient supply was developed recently by wells in the northeast part of the district to carry on extensive placer operations.

Climate and vegetation

Hillsboro, at an altitude of 5,236 feet, has a mean annual maximum temperature of 73.1°F. and a mean annual minimum temperature of 43.0°F., the annual average being 58.0°F. The annual precipitation averages 12.21 inches, most of which occurs in July and August. During these months cloudbursts are common and occasionally the highways are made impassable by washouts. Light snowfalls are common during the winter, but snow never covers the ground for more than a day or two, as the average annual fall is only about 10 inches.

The vegetation is sparse and consists mostly of greasewood and similar desert shrubs and bushes. In the valleys and low places, where the water table is near the surface, willows and cottonwood trees abound. Cedar and piñon trees are seen occasionally on the hillsides, and the alluvial planes to the east of the district have enough grass to support a few cattle for a few months during the year.

Topography

The Hillsboro district is in the foothills on the east slope of the Black Mountains. The elevation ranges from 5,000 feet at the placer deposits in the eastern part of the district to 6,100 feet on the highest peaks and the high slopes in the northwest part. The general nature of the topography is rough and hilly. The steepest slopes range from 15° to 20°, except on some of the highest peaks or in the deep ravines, where they may be as steep as 25° to 30°.

The alluvial deposits to the east of the foothills are comparatively level, except where they have been cut by gulches.

Geology[5]

The rocks of the Hillsboro district consist principally of andesite and latite flows and monzonite intrusions. Small patches of rhyolite and rhyolite tuff indicate that there was at one time a very extensive rhyolite flow probably covering the whole area to a depth of several hundred feet. This flow was removed subsequently by erosion. The fragments from this rhyolite form the principal part of the lower ends of the alluvial deposits

---

[5] Harley, George Townsend, The Geology and Ore Deposits of Sierra County, N. Mex.: New Mexico Bureau of Mines Bull. 10.

- 42 -

to the east of the area. The main igneous area, which is of Tertiary age, is flanked on the northeast and the southwest by shales and limestone of the lower Paleozoic age. Faulting along the northeast and southwest indicate a relative downthrow of the igneous area. Limestone was encountered in a drill hole from the bottom level of the Rattlesnake mine at a depth of 1,100 feet below the collar of the Rattlesnake shaft. The relative downthrow is estimated to be about 1,400 feet owing to the position of similar limestone formations south of the fault.

The sequence of geological activity was briefly as follows: (1) Faulting first took place in the Paleozoic shales and limestones with a subsidence of the area now occupied by igneous rocks; (2) lavas consisting of latite, andesite and rhyolite were exuded and covered the whole depressed area between the faulting on the northeast and southwest. General subsidence may have occurred during or even after the time that the lavas were exuded; (3) intrusion of monzonite porphyry took place just north of the center of the depressed area and in the southwest along the fault plane; (4) latite dikes were formed in the andesite and latite flows. These dikes probably have their origin in the deep-seated source of the monzonite stocks, as they show a tendency to radiate from the larger of two. They later became the seat of the most productive veins of the district. Rich pay streaks that vary in width from 3 to 6 inches may be on either side or within a dike. There are other types of deposits, consisting of disseminated deposits in shear zones in the monzonite and replacement deposits in limestone, but these have been of little commercial importance.

The gold placer deposits of the district consist principally of alluvial fans formed from the erosion of the flows and intrusions of the igneous area. The lower parts of these fans consist mostly of rhyolite fragments and contain very little gold. This would indicate that the rhyolite flows that once capped the area to the west contained few mineral deposits. The upper part of the fans consists mostly of andesite, latite, and monzonite fragments and contain most of the gold. The valuable part of the deposit is roughly separated from the valueless part below by a false bedrock of caliche about 2 feet thick. After the fans were formed they were eroded by streams cutting through them and the caliche is exposed in many of the gulches. During this erosion, which is still going on, much of the higher-grade gravel was washed away and redeposited farther down the gulch, where it is found in irregular runs.

Deposits in Wicks gulch, Rattlesnake gulch, and Bonanza gulch, although containing much less detrital material than the alluvial fans to north and east, have produced some rich placers, from which it is said several small fortunes were derived.

The principal fan in the district was formed from drainage to the east from the main monzonite stock through Grayback gulch and Dutch gulch.

History[6]

Detailed information on the discovery of ore in the Hillsboro district is not available. The first discoveries were probably made about 1875, as immediately after Sierra County became the principal producer of gold in New Mexico.

Production

The value of the metal production of the district from 1877 to 1931, inclusive, was as follows:

Placers:

Rattle Snake placers........................	$40,000
Wicks Gulch...............................	100,000
Luxemburg................................	2,000,000
Miscellaneous.............................	60,000
Total placers...................................	$2,200,000

Lode mines:

Empire Bickford...........................	1,000
Garfield-Butler...........................	40,000
Bigelow..................................	5,000
Mary Richmond group.......................	600,000
Bonanza group............................	700,000
Rattle Snake group........................	1,500,000
Opportunity group.........................	670,000
McKinley-Sherman-Caballero................	21,000
Ready Pay mine............................	10,000
Wicks group..............................	150,000
85 Mines................................	5,000
Chance-Christmas-Feeder Extension........	6,000
Stemburg-Copper King group..............	8,000
Happy Jack...............................	6,000
Tripp-Homestake..........................	50,000
El Oro-Andrews...........................	200,000
Miscellaneous.............................	728,000
Total lode mines.............................	4,700,000
Total placers and lode mines.................	6,900,000

In July 1936, production from lode mines consisted of an occasional shipment of crude ore from old properties where rehabilitation and development work was being done and about 4 tons of high-grade sulphides were chipped from the Bonanza group.

---

[6] Harley, work cited.

Lode Mines[1]

General

The mine workings in the district are fairly extensive but none of them are deep enough to have encountered serious underground water difficulties. The deepest shafts are the Richmond, the Rattlesnake, the Opportunity, the Wicks and the El Oro, ranging in depth from 500 to 525 feet. The Bonanza mine, developed by tunnels, has attained a depth of 500 feet under the highest point of the hill.

Mining throughout the district was by a modified cut-and-fill method of stoping. The high-grade ore was sorted and run through ore passes to chutes in the drift below. The waste was used for filling in the stopes. Where the waste was not sufficient to fill the stopes, wall rock was broken or the stopes were left open.

Development work consisted of drifting along the veins on each level and following the high-grade streaks and drifting along the walls of the veins where the ore pinched out.

Wicks Mine

The Wicks mine is in Wicks Gulch, about 4 miles northeast of Hillsboro and about 3/4 mile west of the main highway from Hillsboro to Caballero. The property is held in trust by Walter K. Mallette, of Spokane, Wash., and is operated under lease by A. A. Luck, of Hillsboro.

Development consists of an adit 700 feet long, two incline shafts on the vein, one 525 feet deep and one 360 feet deep, and three levels from the shafts. No. 1 level is 800 feet long and is connected with the 520-foot shaft only. No. 2 level is 1,200 feet long and is connected with both shafts. No. 3 level is 700 feet long and is connected with both shafts.

The Wicks vein bears N. 15° E. at the Wicks mine, but farther to the south it bears due north and south. The dip is about 70° W. It is from 4 to 7 feet wide, with a pay streak ranging in width from 3 to 6 inches containing quartz, pyrite, bornite, chalcopyrite, and free gold. Hand-sorted ore is said to range in value from $50 to $200 a ton. Total production probably has been about $150,000.

Present operations consist of cleaning out and retimbering the shaft and preparing the levels for production. From 12 to 17 men are employed. Muckers are paid $2 a day and miners $3 a day.

Equipment consists of a gasoline hoist, gasoline compressor, jackhammer, stope, and blacksmith shop. Conventional bits are used. Water is supplied by bailing from the shaft.

---

[1] Harley, work cited.

Bonanza Mine

Situation. — The Bonanza mine is 3 miles north of Hillsboro and 2 miles north of the main highway from Hillsboro to Caballero.

The property consists of five patented claims owned by the Peterson Estate of Sweden and is held in trust by Miller Brothers of Kansas City and Edward D. Tilman of Hillsboro. It is operated under lease by the Las Animas Development Corporation, J. H. Brown, president, 29 South Street, Chicago.

History. — The history and early production statistics of the mine are incomplete and unreliable. A mill was constructed in 1904 and began operating in December of that year. From December 1904 until September 1905, the net returns of the mine were $7,099 from shipping ore, $4,821 from concentrates, and $23,801 from bullion, a total of $35,721. At first the mill had 10 stamps, but the number was soon increased to 20. The returns from three carloads of hand-sorted ore shipped in 1905 showed the following assays: Gold 8.05 ounces per ton, silver 32.3 ounces per ton, and copper 14 percent. It is estimated that the total production to date is about $700,000.

In 1931, development work was done by the Colorado-New Mexico Gold Mining and Milling Co., which opened the winze below the tunnel level and drifted north and south. Stoping was done for 200 feet along the drift to a height of about 30 feet. In 1932, the old mill was remodeled.

The mine was retimbered and dewatered recently and in July was reported to be ready to go on a producing basis. Up to that time production had been limited to about 2 tons of high-grade sulphide a month. From 10 to 15 men were employed, but plans were being made to increase this to 25 or 30.

Geology. — The vein strikes N. 37° E. and dips 80° to 90° northwest. It is from 2 to 8 feet wide and rather irregular. In some places it resembles a fracture zone and in others a shear zone with the ore streaks arranged in lenses that cut obliquely across the general strike of the vein.

The upper portion of the vein is oxidized and contains free milling gold, but the zone of oxidation and enrichment is very irregular. Considerable free gold was mined from the lower level, and sulphide ores were mined from the upper and intermediate levels. The minerals in the ore of the oxidized zone are chalcocite, argentite, limonite, calcite, and silica. In the shallow workings, free gold is found associated with limonite and in vuggy quartz, from which pyrite and chalcopyrite crystals were dissolved. The primary ore consists of pyrite, chalcopyrite, and small amounts of sphalerite and galena in a gangue of quartz and calcite.

Development. -- The vein has been prospected by three adit levels.
There is a total of 6,500 feet of tunnels and shafts. The intermediate
tunnel is 3,500 feet long and extends completely through the mountain.
The lower tunnel is about 2,000 feet long and reaches a depth of about 500
feet below the hilltop.

It is reported that stoping operations were very extensive and continuous, but it is very probable that many low-grade pillars were left.
Much of the stope fill was carefully sampled and it is estimated that by
screening, a milling product can be obtained averaging about $7 a ton.

Costs. -- It is estimated that mining costs above the main tunnel,
largely from old stope fills, would range from $1.50 to $2 a ton; milling
costs would be from $0.50 to $1 a ton; and overhead expenses about 30 cents,
making a total operating cost of $2.30 to $3.30.

Equipment. -- The mine equipment consists of a 450-foot compressor
direct-connected to 60-horsepower gasoline engine, an air hoist, one drifter,
one stoper, and a jackhammer.

Placer Mines

General

Placer operations have been the principal source of gold production in
the Hillsboro district for several years. Limited amounts have always been
produced by sluicing and panning, but the increased production in the past
two years has been due to a number of large-scale operations using portable
washers with Ainlay bowls.

In 1934 and 1935, the John V. Hallett Construction Co. was the largest
producer. In 1935, William Little completed installation of a washing
plant at the Wakely Placer in Gold Run Gulch, 6 miles east of Hillsboro.
The plant consists of a 3/4-yard dragline, a 1-1/4-yard shovel, and a
portable washer with six Ainlay bowls. Seventy thousand yards of material
was handled in 1935, but late in the year the plant was shut down and is
still idle. The Hoot Owl placer was equipped with a dragline excavator
having a daily capacity of 1,000 yards, a portable washer, and a 30-foot
sluice box. The plant was operated only part of 1935.

John V. Hallett Construction Co.

The John V. Hallett Construction Co. of Kansas City is operating the
property of the Animas Consolidated Mines Co. and the Glease lease. This
is a consolidation of the Gold Dust and other placers consisting of about
1,200 acres in Dutch Gulch.

The plant consists of a portable Coulter Ainlay washing plant, two
dragline excavators of 1- and 1-1/4-cubic-yard capacity, a caterpillar
tractor, five storage tanks for water with a total capacity of 180,000

gallons, (reservoir capacity of 1,000,000 gallons), pumping equipment for six wells from 150 to 200 feet deep, and 4 miles of 3- and 6-inch pipe line. All machinery is powered by gasoline engines.

The Coulter-Ainlay washer consists of a hopper of about 2-yards capacity provided with a grizzly for rejecting boulders, a trommel washer that rejects all material larger than 1/4 inch, a stacker for disposing of the tailings, and four 36-inch Ainlay bowls. The washer, the stacker, and the four Ainlay bowls are all operated by a 65-horsepower Buda gasoline engine. All units are mounted on a heavy steel frame on wheels. Figure 10 is a flow sheet of the washer.

The plant is moved from one setting to another by means of a caterpillar tractor. It can be moved a distance of a mile or more and reset for operating in 2 hours.

One dragline excavator is used for stripping and one for supplying material to the washing plant. From 2 to 6 feet of material containing practically no values must be stripped from the surface. Roughly, 50 percent of the total excavating is stripping.

The normal capacity of the plant is about 300 cubic yards in 8 hours. In the event of water shortage, only three of the Ainlay bowls are operated, which reduces the capacity of the plant accordingly. Some water is reclaimed by collecting the run-off in pools and pumping it back to the washer after it becomes clear by settling. According to Mr. Hallett, the greatest operating difficulty is caused by a red sticky clay that is found in certain sections of the deposits. Such sections are unprofitable because of low recoveries and are avoided whenever possible.

The gravel is drawn from the hopper and run over a trommel screen, where water is added. Everything over 1/4 mesh is rejected, and the minus 1/4 mesh is run through the Ainlay bowls. The reject from the Ainlay bowls is pumped to the stacker with a Wilfley pump. The water trickles down from the stacker and is collected in a pool, where it clarifies and is pumped back and re-used.

According to Mr. Hallett, the recovered values amount to 50 to 75 cents a cubic yard for all material run through the washing plant. This checks roughly with the first runs that were made in 1933 by the Animas Consolidated Mines Co. in Dutch Gulch, where the average recovery from 70,000 yards was 39 cents per yard with gold at $20.67 per ounce. The tailing probably runs around 10 cents a cubic yard.

The cost of handling the gravel runs from 16 to 20 cents a cubic yard, including stripping. Pumping and other costs run from 4 to 5 cents a cubic yard, making a total of 20 to 25 cents a cubic yard. This does not include capital charges nor payment of royalties.

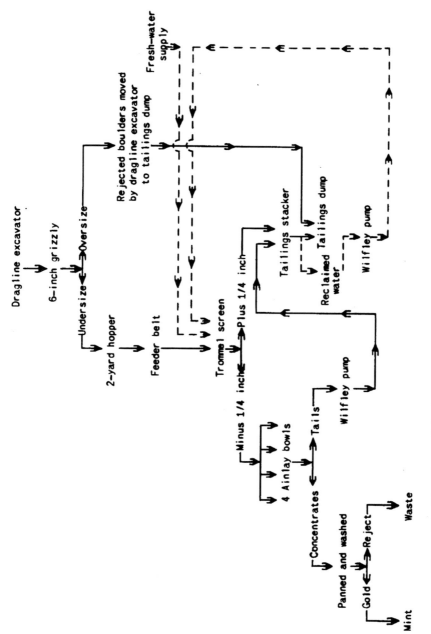

Figure 10.— Flow sheet of Coulter Ainlay washer used by John V. Hallett Construction Co.

The labor required to run the plant one shift a day was as follows:

1 foreman	$250	per month, or	$8.35	per day
3 dragline operators	75	cents per hour	18.00	do.
1 mechanic	60	do.	4.80	do.
1 operator and washer	75	do.	6.00	do.
1 oiler	40	do.	3.20	do.
2 laborers	40	do.	6.40	do.
1 pumpman	40	do.	3.20	do.
10			$49.95	

SHANDON OR PITTSBURG DISTRICT[8]

Situation

The Shandon or Pittsburg district is in Sierra County almost due east of Hillsboro and on the east side of the Rio Grande River. Pittsburg is a small settlement at the junction of the river and Apache Gulch, and Shandon is about 3/4 mile north at the junction of the river with Trujillo Gulch. The village of Caballo, with a population of 75, is on the west side of the river on the main highway from Albuquerque to El Paso. Hot Springs, with a population of 1,845, is 16 miles north. Hatch, with a population of 1,021, is the nearest railroad point 25 miles southeast. The main highway from Albuquerque to El Paso is on the west side of the Rio Grande River valley at Caballo and is paved north as far as Hot Springs and south to El Paso.

Topography

The elevation at the Rio Grande River is about 4,000 feet. The flood plain between the river and the foothills ranges in width from a few hundred feet to half a mile. The alluvial deposits between the flood plain and the Sierra Caballos are from 2 to 2-1/2 miles wide and consist of mesas sloping west toward the Rio Grande River and cut by gulches from the Caballos Mountains that lie just to the east. These mountains are from 5,500 to 6,000 feet high and very rugged.

Geology

The foothills consist of sand and gravel deposits ranging in thickness from 7 to 10 feet at the eastern border near the base of the Caballos Mountains to 60 or 70 feet near the river. Near the eastern border the gravel rests upon a bedrock of rhyolite and rhyolite tuff. In some places andesite is exposed. The gravel near the bedrock is cemented together by manganese oxides and caliche.

The Caballos Mountains consist of a base of pre-Cambrian granite capped by sediments that range in age from Cambrian to Pennsylvanian.

---

[8] Harley, work cited.

The gold-bearing gravels are chiefly in Trujillo Gulch and its tributaries. The gold probably had its origin in the gold-bearing quartz veins in the pre-Cambrian granite that forms the base of the Caballos Mountains. Near the river, where the gravel deposits are thickest, the gold values are said to be too low to work. Farther east, in the Trujillo Gulch, where they have an average depth of about 8 feet, the values are better and considerable mining has been done. In all, there are about 1,270 acres of placer gravels. About one-third of this area is in the gulches and their tributaries and the other two-thirds on the mesas between the gulches.

History

Gold was discovered at Shandon in 1901 and the deposits became known to the public in 1903. During 1904 and 1905 gold production from placers in Sierra County was 1,111 ounces and 2,316 ounces, respectively, most of which came from the placers in Trujillo Gulch. From 1905 until recently the production was unimportant due to lawsuits involving ownership rights. In 1931, legal controversies were settled and the property was taken under a 20-year lease by Caloway and Burke. Considerable sampling was done, and later equipment was installed and operations started.

Warnick Exploration, Ltd.

Situation and General

The original company that started operations in Trujillo Gulch in 1931 was refinanced and reorganized a number of times. At present, the property is being operated by the Warnick Exploration, Ltd., Benjamin C. Warnick, president, Dupont Bldg., Wilmington, Del. The company has title to 1,270 acres of placer ground with the water rights. Operations are about a mile east of the Rio Grande in Trujillo Gulch.

Mining Methods and Costs

Equipment consists of a grizzly for rejecting boulders, a bin or hopper of about 5 cubic yards capacity, steel-lined sluice box, a 1,000-gallon-per-minute, 4-stage centrifugal pump powered by a 200-horsepower Diesel engine, and a 1-3/8-cubic-yard dragline. Water is pumped from the Rio Grande River.

The material is loaded into trucks by the dragline excavator and delivered to the grizzly and sluice boxes. Trucking is done by contract truckers. Two or three trucks are used, depending on the length of haul. The sluice box is about 3-1/2 feet wide and 2 feet deep. Angle irons, steel rail-sections, and other similar material is used for riffles. The clean-up is made at the end of each shift and run through an Ainlay bowl.

A central sluicing plant was chosen in preference to a portable washer on account of the rough nature of the terrain where the operations are being conducted. When moving the excavator to a new location, it is sometimes necessary to stop operations from 1/2 to 2 days for building roads for the trucks. Up to the present time it has been necessary to haul the material from 1/4 to 3/4 miles. In the event of greater distance, it would not be a very great task to move the whole sluicing plant to a more favorable location.

The total operating costs of delivering the material to the sluice box, pumping water for sluicing, and cleaning up are from 20 to 25 cents a cubic yard. The lowest recovered values were 40 cents per cubic yard and the highest $1.02 per cubic yard. The bullion is sent to the United States Mint in Denver.

The plant is operated two shifts a day, the average daily capacity being about 600 cubic yards.

Labor

The labor required to operate the plant on a 2-shift basis, exclusive of trucking, which is done by contractors, is as follows:

Dragline operators	2
Oilers	2
Laborers	6
Pumpmen	2
Total	12

The company furnishes quarters, but the men pay for their board at the company boarding house. The wage scale is relatively low.

PINOS ALTOS DISTRICT

Situation

The Pinos Altos district is in Grant County, about 7 miles north of Silver City.

The town of Pinos Altos, with a population of about 250, is the only settlement in the immediate vicinity. The district is served by a graveled highway and a branch of the Santa Fe Railroad from Silver City. The highway is maintained in good condition, but the railroad has not been in operation for a number of years.

There are a number of old mills and one of more recent construction in the district, but none of them have been in operation in recent years. What little ore is mined is shipped in the crude form to the El Paso smelter. Transportation to Silver City is by truck, and from Silver City to El Paso either by truck or rail but mostly by rail.

Geology[2]

The rocks of the district comprise an intrusion of granodiorite porphyry in a complex of diorite porphyry and associated dykes. Veins were formed in both masses and cut across the contacts without interruption. The diorite porphyry forms the crest of the Pinos Altos Mountains, while the granodiorite is found along the lower eastern slopes.

The important fissure of the district trends from N. 18° E. to N. 30° E. and dips steeply either to the east or west. The outcrops are prominent and in some cases can be traced on the surface for a mile or more. The widths of the veins range from a few inches to 6 feet, generally between firm walls. The vein minerals consist of quartz, pyrite, chalcopyrite, sphalerite, galena, gold, and silver. Barite and rhodochrosite are rarely present.

General Mining Practice

Mining practice throughout the district consists mostly of horizontal cut-and-fill stoping; in a few places veins were encountered of sufficient width and regularity to permit shrinkage stoping. Conditions in most mines favor the horizontal cut-and-fill method of stoping as it provides a means of disposing of waste that must be broken with the ore in narrow veins.

In 1935, shipments were made from six mines and dumps -- the Golden Rule, Golden Giant, Hazard, Kept Woman, Langston, and Savannah Copper Co. The New Mexico Mining Corporation, a subsidiary of the International Mining Corporation of New York City, did extensive development work on about 50 mining claims, including the property of the Savannah Copper Co., but before the end of the year operations ceased and the options were relinquished. Placer gold was produced by individuals engaged in panning and rocking operations. Present activities are limited to two properties.

Hazard Mine

The Hazard mine is about 2-1/2 miles northeast of the village of Pinos Altos. In July 1936 the mine was operated two shifts a day and employed from 12 to 15 men. Two jackhammers were in use, one at development and one at ore breaking. Monthly production consisted of 20 to 25 tons of oxidized gold, silver, copper, and lead-bearing ore, which was shipped to the El Paso smelter.

Equipment consists of a gasoline hoist and a gasoline 2-drill portable compressor.

---

[2] Paige, Sidney, The Ore Deposits Near Pinos Altos, N. Mex.: Geol. Survey Bull. 470, 1910, pp. 109-125.

Golden Rule Mine

The Golden Rule mine is about 1/2 mile north of Pinos Altos and just east of the Pinos Altos Redstone highway. It is being operated under lease by Ed Symons of Silver City. During the summer of 1931 a carload shipment of about 25 tons was made each month. The mine was operated one shift a day and employed four men -- one to run the hoist and three underground. One jackhammer was in use breaking ore.

Equipment consists of a gasoline hoist and a gasoline protable compressor.

STEEPLE ROCK DISTRICT[10]

Situation and Facilities

The Steeple Rock district is in the western part of Grant County near the Arizona-New Mexico line.. It is served by a branch road in fair condition from Duncan, Ariz., 22 miles southwest.

East Camp Group of Claims

Surface buildings and mine equipment were installed on the East Camp group of claims in 1934. Development on this group consists of a 330-foot shaft and drifts on the Davenport claim and a 380-foot adit drift and raises on the McDonald claim. Additional development was done on both of these claims in 1934, and production was started about March 1 and continued throughout the year. The ore was shipped crude to the El Paso smelter.

Production

Available production figures for 1934 and 1935 are as follows:

Year	Tons	Gold, ounces	Silver, ounces	Copper, pounds
1934	1,534	413	20,630	1,700
1935	908	269	14,801	1,285
Total	2,442	682	35,431	2,985

This indicates that the average grade of the ore is as follows: Gold, 0.279 ounce per ton; silver, 14.5 ounces per ton; copper, 0.61 percent.

10/ Henderson, Chas. W., Gold, Silver, Copper, Lead, and Zinc in New Mexico: Minerals Yearbook, 1935, p. 284, and 1936, p. 317.

BURRO MOUNTAIN DISTRICT

Situation

The Burro Mountain district is near Tyrone, 11 miles southwest of
Silver City and about a mile west of the Silver City-Lordsburg highway.

Production

In 1934 and 1935 small amounts of oxidized gold-silver ore were
shipped to the El Paso smelter from the Contact group of claims of the
Phelps Dodge Corporation. The Shamrock group of claims produced a small
tonnage that was treated in a plant equipped with a small ball mill,
amalgamation plates, and concentrating tables. The concentrates were
shipped to the El Paso smelter and the bullion to the Denver Mint.

The entire production of the two groups for 1934 and 1935 was
probably from 300 to 350 ounces of gold. Production continued intermittently during most of 1936.

MT. BALDY DISTRICT

Situation and Facilities

The Mt. Baldy district in the Sangre de Cristo Mountains is in the
western part of Colfax County, about 30 miles south of the Colorado State
line. Elizabethtown, the largest settlement, has a population of about
150.

The district is served by the main highway from Colfax to Taos and
by a branch highway from Therma on the main highway, through Elizabethtown
to Red River and Questa. A branch line of the Santa Fe railroad runs from
Dillon to Ute Park. The nearest railroad point is about 11 miles east of
Therma.

Water Supply

The district is over 10,000 feet above sea level; owing to this high
altitude, there is considerable precipitation. There is no water shortage,
except at some of the placer operations high up on the slopes of Mt. Baldy,
where the streams are small. Ute Creek and other streams furnish an adequate supply for milling.

Geology[11]

Mt. Baldy, the highest peak in the district, with an altitude of
12,400 feet, is made up of faulted sedimentaries intruded by quartz monzonite
porphyry. The igneous rocks occur as numerous dikes and sills intruded into

---

[11] Lee, Willis T., The Aztec Gold Mine, Baldy, N. Mex.: Geol. Survey
Bull. 620, p. 327.

the Pierre shale. On the lower slopes of Mt. Baldy, at an altitude of about 10,500 feet, a basal conglomerate rests unconformably upon the Pierre shale. The gold deposits of the district appear to be very definitely related to the igneous intrusives, but the commercial deposits are mostly on the contact of the conglomerate and the shale. In some cases they are in the conglomerate, but for the most part they are in the shale for a distance of a few inches up to 5 feet from the contact. They are roughly lenticular in shape and comparatively flat, having little more slope than the general contact of the shale and conglomerate. They appear mostly on the down-hill side of anticlinal folds.

History[12]

Copper ore was first discovered in the district between 1860 and 1865. In 1866 placer gold was discovered by a party sent out to do development work on a copper prospect on Mt. Baldy. In the summer of 1867 placer mining began in the region that later became known as the Elizabethtown district. It has been estimated that $2,250,000 was produced by placering, but after a few years it was discontinued as most of the deposits were in gulches high up on the steep slopes of Mt. Baldy, where water was scarce.

Placer gold was found in the draws that came from Mt. Baldy; this led to the search for and discovery of the gold lode that later became the site of the Aztec mine. The discovery was made in June 1868, and by November of the same year development had progressed sufficiently to start production, and a 15-stamp mill had been built and put into operation.

For the first two years the mine produced as much as $21,000 a week, and over a million dollars was taken out by the end of the fourth year. Production previous to 1909 has been estimated to be from $1,250,000 to $1,500,000.

Soon after the discovery of the Aztec mine, the district became prominent, and an English syndicate purchased the Maxwell land grant of 1,700,000 acres, of which the district is a part. The grant was first made by Mexico in 1843, and in 1861 it was confirmed by the United States Congress. By 1872 the ore body of the Aztec mine, which produced such rich returns, was worked out and the mine became involved in litigation. During the 40 years that followed, only spasmodic attempts were made to open new ore bodies, and production in the district was unimportant except for placers. In 1908 and 1909 exploration and development work was started on the Aztec and other mines in the district. The Aztec mill, which at that time consisted of 40 stamps, was remodeled to treat ore by cyanidation as well as by amalgamation and concentration, and by 1911 the lode mines of the district were again producing. In 1914 a bonanza was found in the Aztec mine, and production rose from $63,129 in 1914 to $350,745 in 1915 and $417,258 in 1916. In the 10-year period from 1911 to 1920, production was $1,826,989.

12/ Lee, work cited. p. 325.

Production

The following table shows the production of the Mt. Baldy, Elizabethtown, Therma district from 1909 to 1935:

Year	Placer	Lode
Prior to 1909[1]/	$2,250,000	$1,250,000
1909	16,037	1,705
1910	9,583	3,433
1911	9,860	31,547
1912	8,385	27,316
1913	4,550	15,588
1914	4,930	63,129
1915	2,631	350,745
1916	5,039	417,258
1917	5,739	342,994
1918	2,133	258,034
1919	3,885	238,656
1920	1,634	81,722
1921	3,447	5,361
1922	1,472	76,341
1923	1,811	48,725
1924	2,303	242
1925	970	51,688
1926	1,851	18,745
1927	2,155	32,962
1928	424	18,052
1929	----	----
1930	----	248
1931	103	----
1932	2,430	2,610
1933	1,980	5,196
1934	3,600	36,447
1935	6,220	74,871
Total	2,353,177	3,453,615

1/ Estimated.

Aztec Mine

Recent History

In 1923 the old mill on the Aztec property and a new one that was just completed were destroyed by fire. Another mill was built immediately, and in 1926 it was remodeled. The plant, as remodeled, was of 50-ton daily capacity and consisted of a Universal crusher, 2 Chilian mills, 4 amalgamation plates, 1 cone classifier, 2 Wilfley tables, 2 flotation machines, and 1 Dorr thickening tank. The property belongs to the Maxwell Land Grant Co.

The mine was shut down from 1929 to 1934. In 1932 and 1933 lessees operated the mill on dump ore, and gold was recovered in the form of amalgam and bullion.

In 1934 the mine was reopened and the lessees did development work and opened up enough ore to maintain a steady production of bullion and concentrates. In October 1935 the flow-sheet was changed to flotation. The capacity of the new mill now is 30 tons per 24 hours.

The ore contains free gold and sulphides and arsenides of silver and copper in a gangue of calcite and shale.

Present activities

At present (1936) 60 men are employed by the Maxwell Land Grant Co. doing development and exploration work. Power is supplied by Diesel engines. It is reported that there is some production that is being treated in the flotation mill, but the project is still in the development stage. Most of the work is being done in the Aztec mine, although a number of other mines in the district are on the grant.

The Maxwell Land Grant Co. is owned by interests in Amsterdam, Holland. J. van Houten, of Raton, N. Mex., is the United States agent and attorney.

Montezuma Mine

The Montezuma mine belongs to the Maxwell Land Grant Co. It was re-opened in 1933 by lessees. The ore is treated in a small amalgamation and concentration mill at the mine. The concentrates are shipped to the Golden Cycle mill in Colorado Springs and the bullion to the United States mint at Denver.

Rebel Chief Mine

Production was started at the Rebel Chief in 1934. Small amounts of soft muddy ore are treated in a hand washer. Hand sorted high-grade ore is ground in a hand mortar before treatment.

Red Bandana Mine

The Red Bandana is privately owned and operated. Small lots of crude ore were shipped to the Golden Cycle mill in Colorado Springs in 1934. Since 1935 the ore was ground in a small Huntington mill and run over copper amalgam plates.

Ajax Mine

In 1935 the Ajax Mining Co. shipped 200 tons of gold ore containing 207 ounces of gold to the Golden Cycle mill at Colorado Springs from the Golden Ajax mine on Willow Creek.

Placer Mines

In 1935 the production from placers was $6,220. Most of this was from hydraulicking operations and drift mining on South Ponil, Ute, and Willow Creeks.

During 1935 a testing operation was carried on by the New Mexico Gold Producers Corporation on its lease from the Maxwell Land Grant Co. One hundred forty-six thousand yards of material were handled by a dragline excavator and treated in a flotation and washing plant. The experiment was unsuccessful.

LINCOLN COUNTY

Situation and Facilities

Lincoln County is in approximately the center of New Mexico, 75 to 100 miles east of the Rio Grande River. The southwestern part of the county is rugged and mountainous. The Jicarilla, Captain, Tres Cerro, White, Mescalero, and Sacramento Mountains form a range that extends south through Otero County to the Texas State line.

The county is served by two U. S. highways. U. S. No. 71, in the western part of the county, runs north and south from Corona, in Lincoln County, to Alamogordo, in Otero County. The one in the southern part of the county, U. S. No. 47 and 49, runs east and west from Roswell, in Chavez County, to Carrizozo, in Lincoln County.

History

The gold production of the county was never very impressive. The largest annual production was in 1914, when 3,060 ounces was produced. A number of lode mines were developed, and mills were built to treat the ore, but very few of these ventures were successful. Placer mining accounted for a substantial but minor proportion of the total production.

The Homestake mine of the White Oaks district was the largest and steadiest producer in the county. The property was taken over by the Wild Cat Leasing Co. in 1906 or 1907 and put into production. The ore was milled in a 20-stamp amalgamation mill. For many years it was the principal producing mine of the county. In 1915, the White Oaks Mines, Consolidated, acquired ownership. Production fell to less than half of what it had been up to that time, but it was kept going until 1921, when it was shut down.

Other old mines in the county are the Murphy group in the Jicarilla district, and the Parsons, Hopeful, and American mines in the Nogal district. The Murphy group was operated by the Wisconsin Mining and Smelting Co. from about 1906 until 1910. In 1908 and 1909 large tonnages were treated in a 40-ton amalgamation mill. The Parsons, Hopeful, and American mines were operated by the Parsons Mining Co. in 1914, 1915, 1916, and 1917. A 250-ton cyanide mill was completed in 1917, but little if any ore was ever treated

Most of the placer gold from the county came from the Jicarilla district. It furnished a small but steady output up until 1925, when gold production in Lincoln County ceased entirely, and did not revive until 1932. Gold production from 1932 to 1935 was as follows:

Ounces

1932..............	478
1933..............	703
1934..............	1,042
1935..............	893

The present gold production of Lincoln County is almost entirely from the Jicarilla, the Nogal, and the White Oaks districts.

Jicarilla District

The Jicarilla district is near the south end of the Jicarilla Mountains. The village of Jicarilla is the nearest settlement. It is a few miles to the southwest of the district and has a population of about 275. The district is served by a branch highway running from Ancho, on U. S. Highway 71, southeast to Captain, on U. S. Highway 47.

Production is mostly from placers. In 1933, 50 tons of ore was shipped to the smelter at El Paso and 80 was treated in a 10-ton amalgamation plant at the Lucky Strike mine. The amalgamation equipment at this mine was moved out of the district later.

Placer operations consist of panning, sluicing, hand rocking, and work with Ainlay bowls. Most of the recovered gold is sold to local dealers in Ancho and Carrizozo but some is sent directly to the Denver Mint. In 1935 a power shovel and other equipment for large-scale operations was moved into the district, but it was operated for only a short time.

Nogal District

The Nogal district is on U. S. Highway No. 47, 13 miles east of Carrizozo. Nogal, the nearest settlement, has a population of about 200.

The Helen Rae mine is the principal producer of this district. This mine was first mentioned in the Minerals Yearbook in 1920. From that year until about 1925 considerable development work was done but only occasional shipments of bullion or concentrates were made. The mine is operated by the Helen Rae Mining Co. Shipments of bullion were sent to the Denver Mint every year since 1932. The ore is treated in a small amalgamation mill consisting of a small crusher and a Huntington mill. Other production is from small-scale placer operations.

White Oaks District

The White Oaks district is approximately on a straight line from the Nogal to the Jicarilla district and about half way between. It is about 8 miles east of U. S. Highway No. 71 and is served by a branch highway running southwest from the village of Jicarilla and joining highway No. 71 just southwest of the district. The village of White Oaks, with a population of about 100, is the nearest settlement.

The El Aviador Gold Mining Co., operating the Smuggler and Little Nell mines, is the principal producer. Some of the ore is shipped to the smelter at El Paso and some is treated in an amalgamation concentration mill of 15 tons daily capacity at the Little Mack mine. In 1935, 612 tons of ore containing 279 ounces of gold and 153 ounces of silver was shipped to El Paso, and 117 tons was treated at the amalgamation mill, yielding 93 ounces of bullion 835 fine.

SOCORRO COUNTY

Production

Gold production of Socorro County is principally from the San Mateo Mountains, the Silver Hills district, and the Rosedale district. From 1917 until 1934 the total annual gold production of the county was negligible; in 1934 it was 161 ounces and in 1935, 550 ounces.

Magdalena District

The Magdalena district produced small amounts of gold from lead, zinc, and copper ores. From 1913 to 1917 the production from this source was from 100 to 200 ounces a year, but since then it has never been more than 5 to 10 ounces a year, except in 1925, when it was about 55 ounces. In 1932 the Kelley Magdalena mine of the Empire Zinc Co. made shipments of lead-zinc ore containing gold and silver. Since then the production of the district has been negligible.

Silver Hills District

Production from the Silver Hills district is mostly from test runs made at a testing plant of 25-ton daily capacity. This plant was started in 1933 and used flotation. It continued operating in 1933 and 1934, but in 1935 flotation was replaced by a 20-mesh screen and a sand table and tailing launders lined with cotton blankets. Production in 1935 was 243 pounds of concentrates containing 2.30 ounces of gold.

San Mateo Mountains

In 1934 the Springtime Mining Co. and T. B. Everhart began making shipments from the San Mateo Mountains. Both of these operators shipped from newly developed claims. In 1935 the Springtime Mining Co. completed a flotation mill of 40 tons daily capacity and treated 1,295 tons of ore from the Panky mine. The concentrates produced contained 280 ounces gold and 4,800 ounces silver. Development in the mine consists of an adit 600 feet long and a shaft 100 feet deep with levels at 50 and 100 feet from the collar. Other operators produced a 35-ton shipment of gold-silver ore that was sent to El Paso.

Rosedale Gold Mines Limited

Situation

Rosedale gold mine is situated on the northeastern slope of the San Mateo Mountains in the County of Socorro, New Mexico. The nearest town is Magdalena, on the Santa Fe Railroad, from which point a first-class gravel road runs 29 miles southwest to the mine. The mine is at an elevation of 7,500 feet in the heart of a pine forest, and the camp enjoys an exceptionally temperate climate.

History[13]

Gold was discovered in the San Mateo area about 1882, and the Rosedale property was operated almost continuously from 1886 to 1911. Reports state that dividends were paid for about 15 years simply from ore taken from development. The property was reopened in 1913 and ran until 1916, when a fire destroyed the mill and surface plant. No data are available as to actual production, but it is evident that during early days the mine produced several hundred thousand dollars worth of gold.

In 1934 the present operators obtained the property, reconditioned the mine, and erected a complete cyanide plant of 65-ton daily capacity. After various test runs, the plant went into steady operation in November 1935 and was operated continuously until November 1936, when it was shut down for 3-1/2 months to permit enlargement of the power plant, increase in the mill capacity, and straightening one or two reverse bends in the main shaft. Early in March 1937 the plant resumed production, and at present the mill is treating 135 tons of mine ore daily.

Topography and geology

The San Mateo Mountains cover an area of some 800 square miles, extending for about 45 miles in a southwesterly direction. The rocks of these mountains are an accumulation of volcanics, chiefly rhyolite flows. No sediments occur, except in the southern part of the range at Nogal Creek, where a patch of Magdalena limestone (Pennsylvanian age) is reported.

The rhyolite in the Rosedale area is of Tertiary age and the district is remarkably free from dykes or other types of intrusives.

---

[13] Lasky, Samuel G., The Ore Deposits of Socorro County, New Mexico: New Mexico Bureau of Mines Bull. 8, 1932, pp. 94-95.

The Rosedale vein occurs in a prominent shear zone in a hard rhyolite prophyry. This shear is a main structural feature in the area and can be traced for at least a mile to the southeast of the shaft and for a similar distance to the northwest, as the hard porphyry forming the wall rock of the vein withstands weathering action to form rugged ridges on the slopes and along the backs of the hills. The shear containing the vein has an average strike of N. 15 W. and a dip of 75° to 85° to the southwest.

The shearing extends for a considerable distance into the wall rocks, and for this reason the vein walls are loose in many places and not well defined. The vein filling is mostly a white quartz, crushed and broken portions of rhyolite recemented by quartz to form a type of rhyolite breccia, some limonite, a little hematite, and in places much manganese, which occurs as a black film on the fractures. The gold is free and is found as minute films on the quartz and rhyolite fragments. Usually, where the vein is rich, a considerable quantity of greenish quartz is present, which has a waxlike luster, and this type of quartz is one of the markers for high-grade ore. Oxidation has been intense and extends to the 700 level at least. No sulphides have ever been detected, and although water is present immediately below the 700-level station, it seems likely that the oxidation zone will extend to a much greater depth.

The gold values are spotty. Ordinarily, the values will lie in a high-grade streak; at times a rhyolite breccia band may carry the values, and again a streak of banded quartz may carry high gold values. Such streaks cannot be relied upon to carry for any great distance, as the high-grade portion may suddenly jump from one wall to the opposite side. It is possible and very probable that as the mine is deepened, and especially as the bottom of the oxidation zone is reached, the vein values will be more consistent.

Some premineral faulting is evident, but it is due to the great amount of post-mineral faulting that the ore zones have been cut into blocks and moved considerable distances. Considerable geologic study is being given to the vein structure, throw of faults, etc., but a great additional amount of such work must be done before a complete story can be written as to the movements of the Rosedale vein. Briefly, one or more major post-mineral faults has shifted the vein to the south and at the same time depressed or forced the ore blocks downward. In places, large ore blocks may be found in the heart of such fault movement, but it appears that the larger ore zones are outside the areas of greatest fault disturbance. The faulting has caused a thickening and thinning of the vein, and where thickening occurs the better grade ore is found. Ore blocks will average 100 to 150 feet in length and 100 to 200 feet in height, and in such areas the vein will average 5 to 15 feet in width.

In many places the vein walls are not well defined and the ore is mined to an assay wall.

Development

The mine was opened by a 2-compartment shaft sunk on the footwall of the vein to a depth of 726 feet. When the mine was first worked, levels were driven north and south from the shaft in the vein at 50-foot intervals. For present operations, the level interval has been increased to 100 feet.

Prospecting

The ore zones are prospected by driving drifts and crosscuts, then blocked out between the levels by means of 2-compartment raises driven 100 to 150 feet apart. Such raises afterwards serve as entrance to the stopes and provide proper ventilation to the stoping areas.

Mining methods

The 2-compartment raises have an untimbered chute compartment and a manway compartment closely cribbed with 3- by 12-inch cribbing 4 feet long. The manways have a timber slide for lowering drill steel and supplies to the stopes, a ladderway, and air and pipe lines.

Two men — a miner and a helper — are used in driving raises.

Drifts are driven 5 by 7 feet in the clear, no timber being required. On drift work, one man operates a rock drill, a second man being required only to aid in blasting.

Each level block between the completed raises is prepared for stoping by installing chutes at 15-foot intervals with rock pillars between. The first round for each chute pocket is blasted on the track. The 6- by 8-inch chute timbers are then put in place and a chute of 3-inch lagging is built. Light rounds are then drilled and blasted into the chute. The chute pockets above the chutes are enlarged until they are connected, leaving a pillar from 8 to 10 feet thick between the chutes and a level back 20 feet above the rail from which to start stoping.

This practice has proved to be fast and safe and it does not tie up a level with large piles of muck on the track.

Practically all stoping is by the shrinkage method. Excessive dilution from loose wall rock is prevented by keeping the stopes as full as possible while breaking is in progress. Each cut or slice is carried the full length of the stope before the next one is started, thus preventing the exposure of more than 10 feet of wall rock between the top of the broken ore and the back of the stope. The ore does not tend to break in large pieces. Close attention is given to the placing and loading of drill holes; hence, very little chute blasting is necessary and very few boulders are sledged on the level grizzlies. Breakage in the stopes averages 49.3 tons per machine shift.

An 8-foot pillar is left between the back of the stope and the level above. After a stope is finished, the air and water lines are removed from the raises.

When drawing a stope, the top of the ore pile is kept level by close drawing control. This prevents further dilution while the broken ore is being extracted. Loose rock from the walls falls on top of the broken ore and does not appear in the chute until after a large percentage of the ore has been extracted.

Very few pillars are mined after the stopes are once finished. Waste from development is used as filling in old stopes; however, stope refilling is unnecessary, as not enough ground has been opened up to cause rock movement.

The ore is hand-trammed in 16-cubic-foot standard box-type mine cars and dumped into shaft pockets over 8-inch grizzlies. The skip tender loads from hand-operated chutes to a 1-ton skip cartridge, which discharges direct into a 1-ton skip. The skip dumps automatically into the headframe ore bin.

Worthington 86 drifters mounted on 3-inch columns are used in all drifts, crosscuts, and stopes. Stoping machines are used in the raises. One-inch, round, hollow-lugged steel is used in the drifters and 1-inch, hollow, quarter-octagon steel in the stopers.

One- and one-eighth-inch 40-percent gelatin is used in all cut holes in the drifts and raises. All other holes are blasted with 1-1/8-inch Gelex powder.

Channel samples are taken from every face after each blast in order to maintain close control on vein values. The engineer supervises all sampling and development work and has charge of all mine records.

All timber is cut to standard size in the mine timber yard. Native-pine timber is used throughout the mine.

The mine is cool and there is no dust. There is no water above the 700 level. Good ventilation is maintained in all working places. All working-shaft stations and ore-loading stations are electrically lighted. Mine refuse is removed from the mine daily.

The mine operates two 8-hour shifts for 6 days a week. On Sunday, usually one haulage and hoisting shift is required to maintain sufficient ore for the mill. The mine employs only white labor of the highest-skilled type. Lately, due to the country-wide shortage of skilled labor, it has been necessary to train a number of local men. Many of these men are turning into first-class machinemen and miners, although they had never worked underground before.

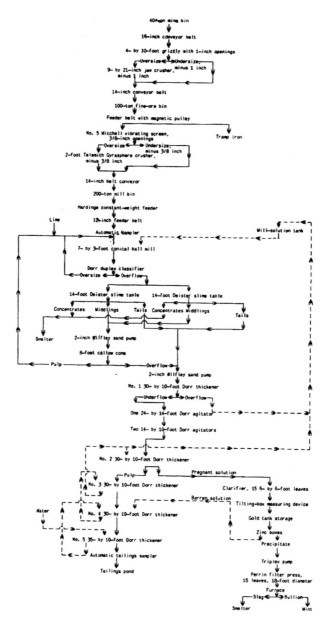

Figure 11.- Flow sheet of Rosedale Gold Mines, Ltd., mill.

During the last three months bonus and contract systems have been introduced. Drifts and raises are run on a contract basis at a published price per foot of advance. Powder, caps, and fuse are sold to the contractors at cost price.

Stopes are bonused on a certain price per cubic yard of extraction. Bonus rates may be changed as conditions demand, but new rates are posted at least a week before the end of each period. Stopes are measured at least twice a month by the mine engineer, and bonus and contract sheets are posted before the following pay day. All miners, trammers, and muckers on bonus or contract are guaranteed the regular rate of daily pay for the type of work performed. A total of 23 men is employed on the mine force, which includes foremen, miners, trammers, muckers, and hoistmen.

Underground men wear the usual type safety hats. Safety kits are maintained on all levels and safety rules are rigidly enforced at all times. Since July 1935, when the first underground work commenced, there have been no serious accidents in the mine and only a few accidents of a minor nature, such as cut fingers, etc.

Milling

Figure 11 is a flow sheet of the Rosedale mill.

Ore from the 40-ton head-frame bin is fed directly to a 16-inch conveyor belt, discharging over grizzlies with 1-inch openings to a 9- by 21-inch Telsmith jaw crusher set to 1-inch opening. The undersize from the grizzly and the discharge from the crusher feed onto a 14-inch conveyor belt, which feeds to a 100-ton fine-ore bin. Ore from the fine-ore bin discharges onto a feeder belt equipped with a Dings magnetic pulley to remove tramp iron, and thence onto a No. 5 Mitchell vibrating screen with 3/8-inch openings. The oversize from the vibrating screen feeds direct to a 2-foot Telsmith Gyrasphere crusher, which crushes to minus 3/8 inch. The screen undersize and gyrasphere discharge fall on to a 14-inch conveyor belt discharging into a 200-ton mill bin. Provision has been made to close-circuit the gyrasphere crusher and the vibrating screen, but, due to the nature of the rock, little oversize is produced and this installation probably will not be installed until a step-up in tonnage is made.

The ore from the mill bin is fed by a Hardinge constant-weight feeder to a conveyor belt, thence to a 7-foot by 36-inch Hardinge ball mill running in closed circuit with a Dorr duplex classifier. The mill head sample is cut at the point where the feed enters the ball mill by means of an automatic sample cutter, which cuts the entire stream for a portion of the time. This sample falls into a separate bin. Grinding is done in cyanide solution and approximately 55 percent of the gold is dissolved in the ball-mill circuit. Lime is also added in the ball mill from a separate lime tank situated outside the mill building.

Forged steel grinding balls are used. Rosedale ore is extremely abrasive and the liner and ball wear is high. The mill is lined with manganese steel liners, yet a set of liners will grind only about 16,000 tons. A mixture of 25-percent 3-inch and 75-percent 2-inch balls is used. Ball consumption runs 4-1/4 to 4-1/2 pounds per ton of ore. The ball-mill discharge runs at 80-percent density and the classifier overflow 19- to 20-percent density.

The classifier overflow goes to two Deister slime tables. The tables make three products — a concentrate that contains about 10 percent of the total gold produced and sent to the smelter; a middling that is pumped to an 8-foot Callow type cone; and a tailing that constitutes the feed to the cyanide section. The cone overflow joins the table tails to the cyanide tanks, and the spigot discharge is returned direct to the ball mill. Before installing the cone, the table middlings were pumped direct to the classifier, but trouble was experienced, inasmuch as too much water was added to the classifier and a great amount of material ground fine enough for cyaniding polluted the ball-mill circuit. The feed to the cyanide section averages as follows:

	Percent
+ 100 mesh	9
− 100 + 200	26
− 200	65

The table tails are pumped to a 30- by 10-foot Dorr thickener, the overflow is sent to the mill solution tanks, and the underflow to the agitators. The pulp passes through a set of three agitators, thence through a unit of four tanks operated in counter current.

Pregnant gold solution overflows from the tank following the agitators to a vacuum clarifier, thence to a head gold tank, and then through the zinc boxes. Every third day the zinc boxes are cleaned, the sludge being flushed to a triplex pump and pumped direct to a small locked Perrin press. Zinc shavings are cut in the mill on a zinc lathe from zinc sheets purchased direct from rolling mills. The cake in the Perrin press is air-dried, discharged direct into pans, and carried to the smelter, where it is melted into bullion.

Reagent consumption per ton of dried ore is as follows:

	Pounds
Lime	2.87
Cyanide	.50
Zinc	.44

Once each month the mill is shut down for 12 to 16 hours for repairs and inspection. Operating time averages 96 percent.

The mill heads average only a fraction of an ounce in silver and 0.12 ounce per ton in gold. Mill extractions are computed daily. The average recovery is from 90 to 92 percent.

The mill operates three 8-hour shifts 30 days a month. American labor is employed everywhere except in the crushing section. The mill crew consists of 1 mill foreman, 3 operators, 3 mill helpers and 2 crusher-men.

Power plant

Generating equipment consists of two Diesel engines direct-connected to 480-volt generators and having a total output of 365 horsepower. The power plant also includes a 14-1/2- by 8- by 10-inch direct-driven Worthington compressor that supplies air for the mine, and a small, single-stage, Sullivan compressor that supplies low-pressure air for the mill agitators.

Shops

A fully equipped machine shop is maintained on the property. Electric welding equipment as well as oxygen acetylene equipment is maintained, and all repairs except heavy Diesel repairs are made on the property under the direction of a machinist and an electrician.

The steel-sharpening shop is equipped with the usual machines. A fully equipped assay office is run for the proper guidance of the mine and mill operation.

Housing

A bunk house and boarding house are operated for unmarried employees and small residences are provided rent-free for married employees. All residences are electrically lighted and most of the houses are provided with running water. The company employs 48 men, all of whom are housed at the property.

Costs

Detailed costs of all operations are kept at the office. Each day, at noon, the total detailed costs for the previous day and the month-to-date costs are posted. Total operating costs, mine, mill, office, etc., average $2.50 to $2.60 per ton of ore milled.

OTERO COUNTY

The entire gold production of Otero County comes from the Oro Grande district, mostly from placers. In 1934 the owner of the Flying Eagle lode claim did some development work and produced 340 tons of ore averaging 0.10 ounce gold, 5 ounces silver, and 42 percent lead. The ore was not shipped in 1934. The Oro Grande Placer Syndicate is working the Little Joe and Cotton Top placers and has been making regular shipments to the Denver Mint. Individual operators working intermittently at the Center placer, 2-1/2 miles northeast of Oro Grande, recover small amounts with a two-way dry washer run by a gasoline engine.

SANTA FE COUNTY

Production from Santa Fe County is from the Los Corrillos and Golden or San Pedro districts. Some of the production from the Los Cerrillos district is from lode mines but most of it is from placers. Lessees at the Santa Fe mine in the San Pedro district produce small amounts of gold ore that they treat by amalgamation in a 6-foot Huntington mill; bullion is sent to the mint in Denver. Other small operators just east of the village of Golden ship small amounts of crude ore to the El Paso smelter.

GOLD PRODUCTION OF BASE METAL MINES

GENERAL

Up to 1934 over 75 percent of the total gold production of New Mexico was from base-metal mines that produced complex ores containing copper, lead, zinc, gold, and silver. In 1934 the production from base-metal mines was 68-1/2 percent of the total, but in 1935 it was only 52.6 percent of the total.

TERERRO DISTRICT

Pecos Mine of the American Metal Co.

Situation and General

The mine is at Tererro in the western part of San Miguel County on the Pecos River at an altitude of 8,000 feet. It is 17 miles northeast of Santa Fe, the capital of New Mexico, and 14 miles north of Glorieta, the nearest railroad station on the main line of the Atchison, Topeka, and Santa Fe Railroad. A 4-1/2-mile standard-gage spur runs from Glorieta to Alamitos, where the mill is situated. The ore is transported from the mine to the mill over an aerial tramway 12 miles long. There is a good automobile road from the mine to Pecos on U. S. Highway 85.

For many years this mine has been the largest producer of gold in New Mexico. It reached its peak production in 1933, when it produced 19,424 fine ounces, or 73-1/2 percent of the total production of the State. In 1934 production was 15,632 fine ounces, 57.3 percent of the total, and in 1935, 14,816 fine ounces, 44.4 percent of the total.

Geology[14]

The mine is in pre-Cambrian granite overlaid by limestone of carboniferous age. The ore bodies are in a northeast-southwest shear zone from a few feet to several hundred feet thick. The mineralization consists of a mixture of sphalerite, galena, chalcopyrite, and pyrite carrying appreciable amounts of gold and silver.

The ore bodies have been developed for a distance of 2,000 feet along the strike of the shear zone and to a depth of 1,200 feet. They are roughly lenticular in shape and sometimes pinch out abruptly either horizontally or vertically. The lenses usually are disconnected, but the ends often overlap.

Development

The mine is opened by two shafts and two adits. The main shaft is vertical and has four compartments with levels at 100-foot intervals. The bottom of this shaft is 50 feet below the 1,000-foot level. The Evangeline shaft was sunk before the present owners started operating the mine. The old development was done from this shaft. It is a vertical two-compartment shaft about 1,000 feet northeast of the main shaft and is used for handling men and supplies.

Mining Methods[15]

The ore bodies consist of one to four parallel veins separated by bands of soft schist and blocky diorite. Sometimes the veins can be mined separately, but more often the composite mass must be mined and the waste sorted out. The walls of the stope do not hold up and filling must be done as the ore is mined out. In wide stopes the back is held up by temporary pillars of cribbing built from the top of the fill to the back of the stope. The waste is sorted from the ore and left as filling in the stope while the ore is run through cribbed ore passes to chutes in the drifts below. Raises from the back of the stope to the level above are used for running in waste filling, as the waste sorted from the ore is not usually enough to fill the stope completely. Cut-and-fill and square-set-and-fill methods are used. Pillar mining by underhand stoping, similar to the

---

14/ Matson, J. T., and Hoag, C., Mining Practice at the Pecos Mine of the American Metal Co. of New Mexico: Inf. Circ. 6368, Bureau of Mines, September 1930.
15/ Matson and Hoag, work cited.

method used at the Magma mine[16] at Superior, Ariz., has been used successfully with a cost slightly less than square-set-and-fill. The method has
not been put into general use because of the close supervision required.
Normal production of the mine is 600 tons a day.

The mine makes considerable water, due to the proximity of the ore
body to the Pecos River and Willow Creek. Sufficient drainage must be provided before a stope can be started. Total underground-mining cost for
1927, 1928, and 1929, including prospecting and developing, was $4.379 per
ton.

Milling[17]

The ore is transported from the mine to the mill at Alamitos over an
aerial tramway 12 miles long. At the mill, it is treated by selective
flotation methods making a lead concentrate and a zinc concentrate.

A typical analysis of mill heads is gold 0.10 ounce per ton, silver
3.3 ounces per ton, lead 4.9 percent, copper 0.8 percent, and zinc 15.4
percent. Metallurgical results for the year 1930 show that 77 percent of
the gold was recovered in the lead concentrate, 8 percent in the zinc concentrate, and 15 percent was lost in the tails. Of the silver, 63 percent
was recovered in the lead concentrate, 22 percent in the zinc concentrate,
and 15 percent was lost in the tails.

The milling cost for 1930 was $1.069 per ton exclusive of laboratory,
experimental work, and royalties, which were $0.120 per ton, making a total
milling cost of $1.189 per ton.

LORDSBURG DISTRICT

Situation and Facilities

The Lordsburg district, including the Virginia and Pyramid or Shakes-
peare districts, is south of Lordsburg in the Pyramid Mountains in Hidalgo
County about 20 miles by road east of the New Mexico-Arizona State line at
an altitude of 4,237 feet. It is on the main line of the Southern Pacific
Railway and also on U. S. Highway 80, which is paved all the way east to
El Paso and west to within a few miles of the Arizona State line. U. S.
Highway 180, which is also paved, runs northwest from Lordsburg to Duncan,
Ariz.

16/ Snow, Fred W., Mining Methods and Costs at the Magma Mine, Superior,
     Ariz.: Inf. Circ. 6168, Bureau of Mines, September 1929.
17/ Bemis, H. D., Milling Methods and Costs at the Pecos Concentrator
     of the American Metal Co., Tererro, N. Mex.: Inf. Circ. 6605,
     Bureau of Mines, May 1932.

Production

Up to 1931, the Lordsburg district had been a very important producer of gold. In 1929, 1930, and 1931 the production of Hidalgo County was 10,608 ounces, 14,012 ounces, and 11,278 ounces, respectively. At least 90 percent of this was from the Lordsburg district, where the Eighty-Five mine,[18] of the Phelps-Dodge Corporation, was the principal producer.

Eighty-Five Mine

Early in 1932, the Eighty-Five mine was shut down and the gold production of Hidalgo County dropped to 1,412 ounces. In 1933 the production was only 111 ounces, but during the two years that followed a number of small operators started producing, and in 1935 the production was 993 ounces.

Geology

The rocks of the Pyramid Mountains consist of andesite, diorites, and porphyries of similar composition intruded by coarse-grained light-colored granodiorite. The ore deposits are in fissure veins. The ore minerals consist of chalcopyrite and pyrite carrying gold and silver. The gangue minerals in the veins are quartz, calcite, barite, hematite, and rhodocrosite.

Mining methods

Mining at the Eighty-Five mine[19] was by shrinkage and inclined cut-and-fill. If the walls were solid, shrinkage was used. If the walls were loose or if a great deal of waste had to be broken, the inclined cut-and-fill system was used. From 1920 to 1930, 72 percent of the total ore mined was by shrinkage and 28 percent by cut-and-fill methods. Ore that was mined by shrinkage was sorted before it was loaded into railroad cars. Some of the ore that was mined by cut-and-fill methods was shipped without sorting.

The entire output of the mine was shipped in the crude state to the Douglas smelter of the Calumet and Arizona Mining Co. From May 1, 1920, to June 1, 1930, 783,138 tons were shipped. This ore averaged 0.111 ounce gold, 1.23 ounces silver, and 2.79 percent copper.

18/ Youtz, R. B., Mining Methods at the Eighty-Five Mine, Calumet and Arizona Mining Co., Valedon, N. Mex.: Inf. Circ. 6413, Bureau of Mines, March 1931.
19/ Youtz, work cited.

AFTER THIS REPORT HAS SERVED YOUR PURPOSE AND IF YOU HAVE NO FURTHER NEED FOR IT, PLEASE RETURN IT TO THE BUREAU OF MINES. THE USE OF THIS MAILING LABEL TO DO SO WILL BE OFFICIAL BUSINESS AND NO POSTAGE STAMPS WILL BE REQUIRED.

UNITED STATES
DEPARTMENT OF THE INTERIOR
BUREAU OF MINES

PENALTY FOR PRIVATE USE TO AVOID
PAYMENT OF POSTAGE, $300

OFFICIAL BUSINESS

RETURN PENALTY LABEL

THIS LABEL MAY BE USED ONLY FOR
RETURNING OFFICIAL PUBLICATIONS.
THE ADDRESS MUST NOT BE CHANGED

BUREAU OF MINES,

WASHINGTON, D. C.

CPSIA information can be obtained at www.ICGtesting.com
Printed in the USA
BVOW012348231011

274356BV00001B/2/P